Marcio Luiz Varela

Utilisation of kaolin waste in the production of porcelain flooring

Marcio Luiz Varela

Utilisation of kaolin waste in the production of porcelain flooring

Practical application

ScienciaScripts

Imprint

Any brand names and product names mentioned in this book are subject to trademark, brand or patent protection and are trademarks or registered trademarks of their respective holders. The use of brand names, product names, common names, trade names, product descriptions etc. even without a particular marking in this work is in no way to be construed to mean that such names may be regarded as unrestricted in respect of trademark and brand protection legislation and could thus be used by anyone.

Cover image: www.ingimage.com

This book is a translation from the original published under ISBN 978-613-9-72569-4.

Publisher:
Sciencia Scripts
is a trademark of
Dodo Books Indian Ocean Ltd. and OmniScriptum S.R.L publishing group

120 High Road, East Finchley, London, N2 9ED, United Kingdom
Str. Armeneasca 28/1, office 1, Chisinau MD-2012, Republic of Moldova, Europe
Printed at: see last page
ISBN: 978-620-7-91861-4

CONTENTS

SUMMARY

Currently, industries in all sectors are very concerned about the disposal of waste generated during the production process. The mining sector is no different. It was found that the kaolin processing industry generates large volumes of waste. This is because the yield of the kaolin processing process is around 25%, i.e. for every tonne of kaolin processed, 750 kg of waste is generated and only 250 kg of marketable kaolin is produced. This waste is basically made up of kaolinite, muscovite mica and quartz. Disposal of this waste generally has a very negative environmental impact, which has justified research efforts to find a rational solution to this problem. The effect of incorporating waste from the kaolin processing industry on the physical, chemical and technological properties of high-quality ceramic tiles was therefore studied. This study used kaolin processing residue generated by the CAULISE exploration and processing industry, added at levels of 1, 2, 4 and 8% to the standard mix used to produce porcelain tiles from the Cerâmica Elizabeth industry. Both raw materials were characterised using X-ray diffraction (XRD), X-ray fluorescence (XRF), DTA and TG thermal analyses. After sintering the specimens, water absorption (AA), apparent porosity (PA), linear shrinkage after firing (RLQ), apparent specific mass (MEA) and 3-point flexural tensile strength (RTF) tests were carried out to determine the technological properties of these materials. The preliminary results show that the waste studied can be considered a well-behaved raw material with great potential for the ceramic floor and wall tile industry.

CHAPTER 1

INTRODUCTION

The global ceramic tile market has been following a growth trend, with China taking the lead in both production and consumption in this promising market. Brazil's production is growing steadily and with this behaviour it is likely to overtake Italy and Spain in the coming years, becoming the world's second largest producer. China has shown rapid and sharp growth in its exports in the current panorama of the international ceramic tile market. The United States continues to be the largest import market, of which Brazil is already the 3rd largest supplier (ANFACER, 2006). In the domestic market, ceramics have played an important role in the country's economy, with a share of GDP (Gross Domestic Product) estimated at 1%, corresponding to around 6 billion dollars. This is due to the abundance of natural raw materials and alternative energy sources, together with the availability of practical technologies built into industrial equipment. As a result, the combination of these factors has led Brazilian industries to evolve rapidly and many types of products from the various ceramic segments, especially porcelain tiles, have reached world quality levels with a considerable quantity exported (ABCERAM, 2005).

One of the products responsible for Brazil's favourable market situation is porcelain tiles, which have grown considerably in consumption in recent years. Porcelain tiles are a product capable of balancing the production cost/invoice ratio with a much lower production volume than conventional single-fired products and, as they are widely accepted on the foreign market, domestic production of this product is expected to grow rapidly (GILBERTONI, 2005).

In contrast to the rapid development of the ceramic tile sector, there has been a considerable increase in the volume of waste coming from kaolin processing industries. One of the reasons for this increase is that in most cases there is no management plan for the waste produced by its generators, which increases the negative environmental impact and disposal costs. An example of this situation is the amount of waste generated in the kaolin mining and processing industries in Rio Grande do Norte, around 7500 tonnes of waste per month. This is because the yield of the kaolin processing process is around 25%, meaning that for every tonne of kaolin processed, 750 kg of waste is generated and only 250 kg of marketable kaolin. As a result, the disposal of this waste has a negative impact on society as a whole, through the costs of removing and treating it, as well as causing damage to the environment and the health of people living near the dumping sites. Most of the time, this waste is removed and disposed of illegally on land, riverbanks and streets on the outskirts of cities. In this way, the total social cost is practically impossible to determine, as its consequences lead to a deterioration in the quality of urban life in aspects such as transport, flooding, visual pollution, among others. As a result, public authorities, usually town halls, commit resources, which are difficult to measure, to removing and/or treating this waste and the problems it generates.

In view of the environmental, social, health and economic damage resulting from the inadequate disposal of waste generated by industries in the kaolin mining and processing production chain, it is necessary to develop criteria and procedures for its management. This information is essential for the implementation of government guidelines to speed up the process of reducing negative environmental impacts.

Waste management requires the development of processing technologies applied to each case. One rational application for kaolin mining and processing waste is to utilise it in the ceramics industry, as it has the physical and chemical characteristics to be used in the manufacture of ceramic floor and wall tiles, in this case porcelain tiles. This would have positive economic and social effects for the industries that generate waste and for the ceramic tile industry.

These positive aspects are attributed to this type of waste because its constituents are kaolin, quartz and muscovite mica. These phases are of great importance when present in the formulation of high quality ceramic tiles because they give the piece strength in green, in the case of quartz, and for forming the structure, giving strength and forming the liquid phase of the sintered material, in the case of kaolin and muscovite mica, respectively.

An important aspect to consider is that the quality of the products generated must be maintained. Therefore, the products obtained from the use of kaolin processing waste must meet the requirements set out in Standard NBR 13818/1997 - Ceramic tiles for cladding - specifications and test methods.

CHAPTER 2

OBJECTIVE

The aim of this thesis is to evaluate the technical potential of using kaolin waste from its processing as a raw material in the composition of ceramic tiles for the production of high quality porcelain tiles and thus contribute to the development of processing methodologies using waste within the ceramics sector.

This work has the following specific objectives:

1. To study the influence of the addition of kaolin residue on the processing characteristics cf a reference ceramic mass, already produced and used by a consolidated ceramics industry in the market;

2. To study the influence of the addition of kaolin residue on the technological properties and microstructural characteristics of porcelain tile formulations;

3. To help minimise the social and environmental damage caused by the exploration and processing of ceramic raw materials.

CHAPTER 3

PORCELANATE

The name "porcelain stoneware" derives from "stoneware", a name for ceramic materials with a compact structure, characterised by a crystalline phase immersed in a predominant vitreous phase, and "porcelain tile" is the term that refers to the technical characteristics of the product that are substantially reminiscent of porcelain (PALMONARI, 1989). The above definition refers to the commercial name given to porcelain tiles when they first appeared. Technically, there is a classification that makes a clear distinction between what is considered porcelain tile and what is stoneware, i.e. the former belongs to group Bla, which is the group of pressed elements with a water absorption percentage in the range of 0 to 0.5%. Stoneware is in the Blb group, although it is also shaped by pressing, its water absorption percentage is in the range $0.5 > AA < 3.0\%$ (NBR 13818, 1997).

Porcelain tiles can be glazed or unglazed. The former arose from the need to win over the market, where surface aesthetics take precedence over the product's technical characteristics. This means that the glazed product is most applicable to building projects that require this attribute, i.e. the main concern is aesthetics. When the unenamelled type is used, as it was designed to be, its characteristics of extreme surface resistance and a high degree of impermeability are emphasised.

A striking feature is the low degree of water absorption when compared to other types of ceramic tiles, in the order of 0 to 0.5%, as mentioned above. This type of characteristic directly reflects the product's degree of porosity, i.e. the lower the water absorption content, the lower the porosity index and consequently the greater the product's compactness. It therefore has superior mechanical characteristics, i.e. resistance to bending and breaking loads, hardness and resistance to abrasion. Since porcelain tiles are formed by pressing and have low **porosity, their classification, according to Standard "NBR13818 -** Ceramic tiles for cladding **- specification and test methods", is in group Bla,** corresponding to completely glazed ceramic tiles, where the maximum water absorption value is 0.5%.

As far as the dimensional aspect is concerned, during the porcelain tile sintering process, at around 1250 °C, shrinkage occurs in the product, which increases as the temperature rises. This is due to the large amount of liquid phase formed during this process, promoted by the presence of melting materials in the mass. Therefore, this variation should be checked according to the surface area of the piece. As for flexural strength, extreme compaction should be confirmed by the high value of this parameter, which should average 35 MPa or more. When checking abrasion resistance, the two different situations of the product - enamelled and unenamelled - must be taken into account. In the case of the latter, abrasion resistance depends essentially on structural compaction and is therefore greater the lower the porosity. Also in this case, the ceramic tiles must have a removed volume of less than or equal to 175 mm^3 . In the case of enamelled material, the test for determining resistance to surface abrasion, Annex D of NBR-13818/1997 (BIFFI, 2002), can be carried out. According to this standard, the Mohrs hardness value must depend on the specific application and can be checked using available test methods, with the minimum limits being agreed between the parties. From the point of view of linear thermal expansion, resistance to thermal shock, resistance to freezing and resistance to impact, the same procedure, agreed between the parties, must be followed.

Porcelain tiles resemble natural stone, but have numerous characteristics that surpass the performance of marble, granite, São Tomé stone, etc. Porcelain tiles stand out from natural stone in the following respects: greater chemical resistance - suitable for use in laboratories and industries; it is impermeable - greater resistance to stains, easier to clean and in the event of moisture infiltration, there is no development of damp patches; greater resistance to abrasion - recommended for very high traffic areas; uniformity of colours in the piece and between pieces - aesthetic effect pleasing to the eye; lighter, thinner and more mechanically resistant - easier to transport and handle, and finally; easier to lay - a traditional ceramic tile layer can lay porcelain tiles (HECK, 1996).

3.1   TYPES OF PORCELAIN TILES

Porcelain tiles come in a variety of product types, thus reaching the various ceramic tile markets, as can be seen below in a summary classification of the existing commercial types.

Single-colour porcelain tiles are the most aesthetically pleasing products, where pastel colours predominate. They are obtained from atomised powders with uniform colouring. Granite-look porcelain tiles are obtained by mixing differently coloured atomised powders, giving rise to the effect commonly known as "salt and pepper". The background colour is light and shows the natural colour of the basic ceramic mass, which is not coloured. Variegated ceramics are those produced with mixtures of coloured, atomised and also micronised powders which, when properly loaded in the presses, are distributed with a certain randomness, giving rise to superficial and variegated smokiness. Macrogranites are those obtained by mixing atomised powders with large granules, varying in size from 1 to 8 mm, and coloured in different ways. These granules are obtained by regranulation of atomised and pressed powders, which can be coloured internally. The surface of the ceramic tile, which has a colour background similar to that of granite or variegated tiles, as a result of these macro-granules, produces products that are very similar to natural stone.

Products decorated with soluble salts are obtained by screen printing or by dripping (disc, airbrush, etc.) solutions containing chromophoric salts of Fe, Cr, Co, V, Mn, among others. The base of the ceramic is lightly coloured to highlight the decoration, which can be applied to a raw or fired base. Rustic, structured and glazed ceramics are obtained using a pressed ceramic base with structured patterns and enhanced by flaking, serigraphy of glazes or soluble salts and brushing. The ageing effect is very effective and similar to the natural effect of time and use. Stencil stoneware is an application technique, carried out on presses, involving a specific film that results in multicoloured effects on the surface. It is introduced onto the powder to be pressed in the mould cavity before pressing and is fired together with the ceramic body (NOVAES, 1998).

3.2   WEIBULL MODULUS OF PORCELAIN TILES

Ceramic materials have a series of defects that can act as stress concentrators and determine the points where the product begins to fracture. The mechanical strength of a product depends on its microstructure and, above all, on the distribution and size of the defects present. As this distribution is almost always random, the strength evaluated experimentally shows a dispersion. Quantitatively, this dispersion of mechanical strength values can be obtained using the Weibull distribution. The Weibull modulus provides an indication of the reproducibility of the product's mechanical strength. The higher the Weibull modulus, the less dispersed the mechanical strength values. In general, the Weibull moduli obtained for porcelain tiles are quite high, which represents an excellent characteristic of the reproducibility of their mechanical strength. Weibull drew an analogy between a fragile tensile structure and a chain that breaks when the strength of its weakest link is exceeded. The growth of an isolated microcrack in an elastic body, when loaded in the direction normal to its plane, is similar to the rupture of a chain (MENEGAZZO at al., 2002).

3.3 BASIC RAW MATERIALS FOR PRODUCING PORCELAIN TILES

A porcelain tile is basically made up of a mixture of clays, feldspars, feldspathic sands, kaolins and sometimes phyllites and additives when necessary. Feldspars play a melting role in porcelain tiles, as they provide the first liquid phases that appear during firing. These liquid phases bypass the more refractory particles, bringing them closer together through the surface tension forces that are generated in the finer pores, a fact that generates contraction of the piece. In this way, feldspars are initially responsible for the densification process, which contributes most to the densification of the pieces and, consequently, to the desired properties of the porcelain tile. As kaolins are rich in alumina, during firing it can be part of the vitrification reaction forming silica-alumina glasses, or form secondary mullite ($3Al_2O_3.2SiO_2$) in the shape of needles, which act as the body's skeleton, contributing to increased mechanical resistance. The clays, in turn, have the function of providing plasticity, i.e. the ability to mould parts (MENEGAZZO, 2000).

Ceramic raw materials can be classified as plastic or non-plastic. Although both play a role throughout the production process, the plastics are essential in the forming phase, while the non-plastics are more active in the thermal processing phase; ceramics are therefore fundamentally made up of inorganic material, or more specifically, a mixture of oxides. The correct definition of the proportions of the raw materials, as shown in Figure 1, and the physical and chemical characteristics of each of the raw materials used to produce porcelain tiles, is the basis for creating a ceramic mass capable of generating products compatible with the requirements of Standard NBR 13818 - Ceramic tiles for cladding - Specifications and Test Methods. Generally, the raw materials used are from different mineral groups, each of which has a specific influence on the final characteristics of the product (BIFFI, 2002). Therefore, a brief mention of the main raw materials used in the production of porcelain tiles is necessary and can be seen below.

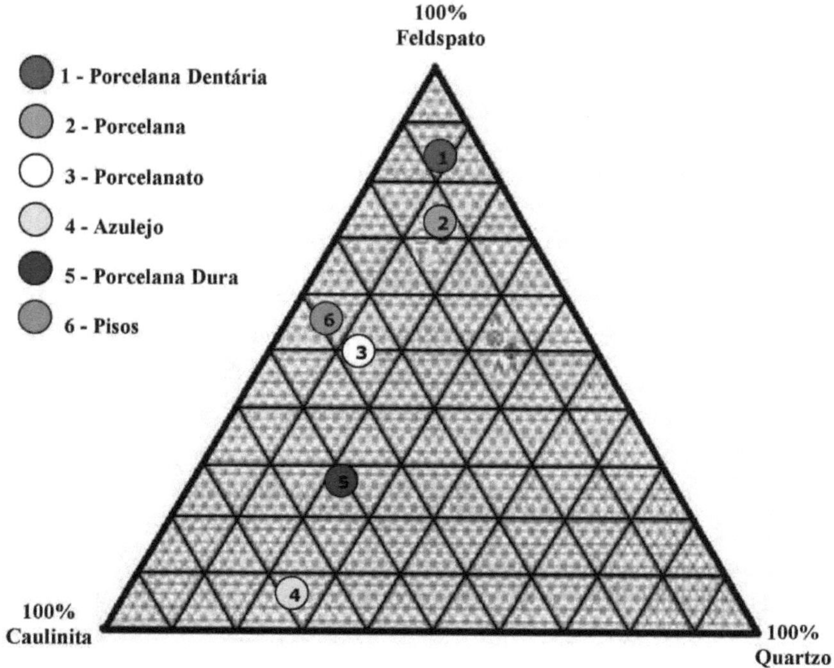

Figure 1 - "Kaolinite - Feldspar - Quartz" ternary diagram

3.3.1 Feldspar

Feldspars, Figures 2 and 3, are the minerals that predominantly give rise to kaolinite and halloysite, in either form, either through weathering or hydrothermal action. In order to be transformed by weathering into kaolinite, a feldspar must pass through the form of muscovite mica as an intermediate phase. To date, there is no experimental evidence of the direct transformation of feldspar into kaolinite without an intermediate phase (SOUZA SANTOS, 1989). The alkalis found in clays are almost entirely due to feldspars, micas or exchangeable cations. They are melting agents and therefore undesirable for refractory materials, but they are essential for vitrifying porcelain and other white ceramic products.

Feldspars, being materials with a high alkali content (Na_2O and K_2O), sodium ($Na_2O.Al_2O_3.6SiO_2$ - albite - Figure 2) and potassium feldspars ($K_2O.Al_2O_3.6SiO_2$ - orthoclase - Figure 3), respectively, when present in the ceramic mass, reduce the firing temperature - as they have a relatively low melting temperature and are therefore used as fluxes or **"glass phase" generators in ceramic masses and glazes** - and the porosity of the product. For the ceramics industry, these two conditions are important for the products, since in addition to lowering energy costs, they reduce water absorption and increase mechanical resistance (ABCERAM, 2005).

Figure 2 - Sodium feldspar - Albite.

Figure 3 - Potassium feldspar - Orthoclase.

3.3.2   Quartz

Quartz (Figures 4a and 4b) is found in nature in polymorphous forms: quartz, tridymite and cristobalite. Phase transformations occur with changes in temperature and each one is given its own specific name. Alpha quartz is stable at room temperature, transforming into the beta variety at 573 °C and into tridymite at 870 °C. At 1470 °C, it transforms into cristobalite until it reaches its melting point at 1713 °C (DEER at al., 1975). The presence of quartz in white ceramics and coating materials is essential as it is one of the components responsible for controlling expansion and adjusting the viscosity of the liquid phase formed during sintering of the ceramic mass, as well as facilitating drying and the release of gases during firing and being an important regulator of the correct ratio between SiO_2 and Al_2O_3 for the formation of mullite ($3Al_2O_3.2SiO_2$). Finely ground quartz can be very useful when mixed with the limestone-containing clays used in the ceramic mass, as above 900°C it reacts with CaO to form calcium silicate and contributes to the product's greater mechanical strength (REED, 1995). An important point to note with ceramic mass containing quartz is that during the sintering of the ceramic body, at around 573 °C, the quartz changes size, with a sudden increase in volume of more than 3%. During this transformation, the heating rate must be slow to avoid the appearance of cracks caused by the sudden change in volume. In the cooling phase, between 1250 °C and 1100 °C, rapid cooling will prevent crystallisation of the cristobalite. Rapid cooling also favours the development of transparent glass, while slow cooling favours the formation of crystals with an opaque effect. Between 573 °C and 300 °C or 200 °C, physical changes occur in the silica, which makes slow cooling advisable at both points (BOSCHI, 2005).

11

Figure 4a - Quartz.

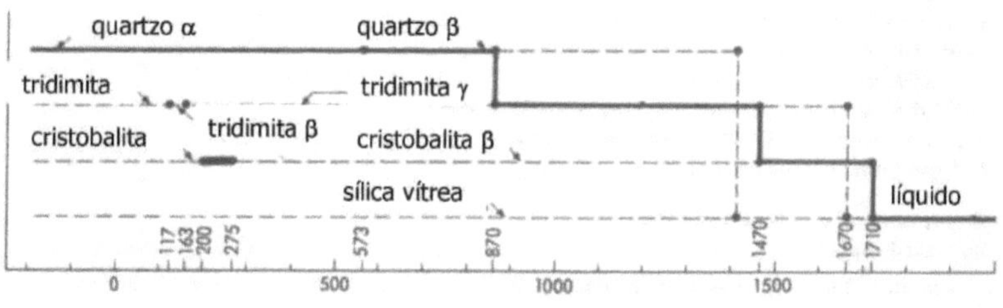

Figure 4b - Diagram of the allotropic transformations of quartz.

(Source: adapted from P.P. Budinikov, 1964)

3.3.3   ARGILA

From a mineralogical point of view, clays are characterised by being largely made up of specific minerals, called clay minerals or clay minerals, with other associated materials and minerals usually occurring in a subordinate way, such as quartz, feldspars, micas, iron and aluminium oxides and hydroxides, carbonates, organic matter, among others (ABREU, 1973).

Due to the presence of clay minerals, clays, when in the presence of water, show a series of properties such as plasticity, wet mechanical resistance, linear drying shrinkage, compaction, thixotropy and viscosity of aqueous suspensions, which explain their wide variety of technological applications. The main groups of clay minerals are kaolinite, illite and smectite or montmorillonite. The basic difference between these clay minerals lies in the type of structure and the substitutions that can occur within the structure of aluminium for magnesium or iron, and silicon for aluminium or iron. The consequence of these substitutions is the neutralisation of the residual charges generated by the differences in the ions' electrical charges by certain cations. Based on this, it can be seen that in kaolinite there is practically no substitution, in illite there is substitution and the neutralising cation is potassium; in montmorillonite there is also substitution and the neutralising cations can be sodium, calcium, potassium and others. This type of occurrence results in the different characteristics of clays, making them suitable for various technological applications.

Clays with a greater presence of specific clay minerals have specific characteristics, such as clays made up essentially of the clay mineral kaolinite, which are the most refractory because they are made up essentially of silica (SiO_2) and alumina (Al_2O_3), while other clays, due to the presence of potassium and iron in greater concentration, have reduced refractoriness. The presence of other minerals, often considered to be impurities, can substantially affect the characteristics of a clay in order to reduce its refractoriness.

The need then arises to eliminate these impurities, either through physical or chemical processes. This process is called beneficiation (ABCERAM, 2005).

Clay for the production of stoneware is generally plastic and is made up of kaolinite and other subordinate clay minerals (illite and smectite), with a variable content of quartz, feldspar, micas and organic matter. In the composition of the mass, they have the function of giving a light colour when fired, providing plasticity and binding characteristics to the ceramic mass, providing rheological properties facilitating fluidity, conferring a good density when fired and giving optimum mechanical resistance to the final products.

The white firing colour is due to the low levels of iron and other colouring elements, as well as considerable levels of kaolinite, a characteristic that makes these clays scarce. Due to this scarcity, for the production of porcelain tiles, mixtures of various types of clay are used, plastic, non-plastic, with a high kaolinite content or richer in clay materials such as illite and montmorillonite, so that the mass achieved confers the physical characteristics necessary for a quality end product (BIFFI, 2002).

3.3.4 CAULIN

Although kaolin is a clay, there will be an item dedicated to it because it is a very important raw material in the production of porcelain tiles. Kaolin is defined as a fine-grained clay, generally white in colour, with good chemical inertness. The most common and industrially important kaolin mineral is kaolinite ($Al_2O_3.2SiO_2.2H_2O$), formed by weathering or hydrothermal alteration. The types of kaolin vary according to their physical characteristics: whiteness, degree of crystallisation, opacity, viscosity, particle shape, etc. The minerals that most commonly make up kaolin, kaolinite, halloysite, diquite and nacrite, have essentially similar chemical compositions, but each of them has structural differences. Kaolin can have coloured impurities, such as hematite, which depreciates its commercial value if it is used to produce paper or light-based ceramic products (PINHEIRO et al., 2007).

Kaolinite is used in porcelain ceramic tiles at levels ranging from 10% to 15% because it gives the tile its own characteristics, i.e. it gives the tile its white colour after sintering. In addition, as it contains a large amount of aluminium oxide (Al_2O_3), it acts as a regulator of the balance of reactions during the vitrification phase of the ceramic mass. Alumina can also take part in silica-aluminium vitreous formation when it is associated with alkaline fusing elements. However, its predominant constitution at the end of firing is as mullite ($3Al_2O_3.2SiO_2$), which due to its structure acts as the "skeleton" of the ceramic materials obtained, thus contributing to increased mechanical resistance (BIFFI, 2002). Ceramic kaolin must have a kaolinite content of between 75 and 85% and no minerals that affect the firing colour, such as hematite (Fe_2O_3), whose content must be less than 0.9%, so that the whiteness index after firing is in the 85 to 92 range (HARBEN, 1995).

3.3.5 OTHER RAW MATERIALS THAT PROMOTE THE FORMATION OF A LIQUID PHASE IN THE DOUGH

Calcite, dolomite, talc, wolastonite, diopside and magnesite are used to reduce the percentage of feldspars in the composition and/or the optimum firing temperature. These elements act as generators of low-temperature eutectics, being minerals based on calcium and magnesia and which can therefore affect colour development as well as the firing range of the clay. For these two reasons, their use in the dough should always be in the lowest possible percentages, and it is inadvisable to use them in percentages greater than 3% (LLORENS et al., 2000).

3.3.6 MICROSTRUCTURE DEVELOPMENT DEPENDING ON THE RAW MATERIALS PRESENT IN THE MASS

During sintering, raw materials containing highly melting alkaline minerals at high temperatures (illite, feldspar, etc.) produce a large quantity of liquid phase, whose viscosity decreases with increasing temperature, causing it to penetrate the existing pores by capillary forces, gradually reducing them (ORTS, 1991). The quartz partially dissolves in the liquid phase and a new crystalline phase, mullite, is formed. The burnt product is made up of a glassy matrix in which mullite particles and quartz particles that have not completely dissolved are dispersed (RADO, 1988; SANCHEZ, 1998). Depending on the characteristics of the raw materials, as well as the clay/feldspar ratio (the main ingredients in the composition), the intensity of the physical-chemical transformations described above can vary considerably, which will lead to significant differences in the properties of the final product. The crystalline phases of porcelain tiles are the same as those found in certain porcelains (SANE, 1951; KOBAYACHI, 1994). In these materials, they are determined by X-ray diffraction (XRD), using a phase diagram to calculate the mullite content. However, porcelain tiles are manufactured in very rapid firing cycles (60 minutes, compared to 24 hours or more for porcelain), which means that mullite cannot be quantified using phase diagrams, which only reflect thermodynamic equilibrium situations. Sintering, on the other hand, is a non-equilibrium thermodynamic process in which a system of particles, aggregates or compacted powders acquires a solid structure after being subjected to a high temperature cycle. The mechanisms responsible for the transport of matter during the sintering process are basically atomic diffusion, plastic creep and viscous creep (BIFFI, 1997).

3.4 PORCELAIN TILE PRODUCTION PROCESS

The composition of the materials used to make porcelain tiles has undergone major changes in recent years. The basic reason for these changes is the technological development that has taken place in the area of ceramic machinery, mills, roller kilns and high-pressure hydraulic presses, all associated with the rapid firing process. It used to take 40 to 50 hours at temperatures of 1200 °C to produce porcelain tiles. Today, with these improvements, cycles of 50 to 70 minutes can be achieved at temperatures ranging from 1200 °C to 1230 °C. The components of the ceramic masses that were subjected to slow firing, due to the long sintering time, had time to participate in the many reactions that led to the total modification of their composition, with the formation of new compounds. In rapid firing, this does not occur, since the components of the mass generally behave with their own characteristics when sintered (SACMI, 1996). A good example of this would be the reduction of kaolin as a raw material which, in rapid firing cycles, behaves as a refractory material, maintaining a high porosity at the end of sintering. On the other hand, in compensation for this situation, there has been an increase in feldspar and plastic clays, which are materials that behave as fluxes. Quartz is always present and its quantity will depend on its degree of purity and the other raw materials. In the case of quartz sands with feldspar, quantities can reach between 25 and 30 per cent, taking into account the fusibility of feldspar (CONTOLI, 1996). Table 1 shows the changes that occur in the porcelain tile mixture.

Table 1 - The evolution of porcelain tile compositions.

Raw materials	1985 - 1990 (%)	From 1990 onwards (%)
Plastic clays	15 a 30	5 a 15
Semi-plastic clays	--	10 a 25
Kaolin	10 a 30	5 a 10
Feldspar	20 a 30	25 a 40
Feldspar sands	5 a 15	10 a 20
Quartz sands	5 a 10	5 a 10
Talc	0 a 5	0 a 3

* Adapted from Grês fine porcellanato - Edizione Sacmi Imola 1996.

The aforementioned changes are aimed at generating a finished product that fulfils the requirements of almost zero water absorption, high flexural and deep abrasion resistance values and low stainability. However, in addition to the changes in the ceramic mass, in order to obtain a product with the typical characteristics of porcelain tiles, the influences caused by the conditions of the production process in the fundamental phases of dosing, milling, pressing, drying and sintering must be taken into account.

3.4.1 CONTROL OF   KEY PARAMETERS

For the production of porcelain tiles, the dosing stages of the raw materials are followed, where the quantity by weight of each component that will make up the ceramic tile is controlled. This control is based on the chemical analyses previously carried out on the raw materials. Grinding aims to homogenise and achieve the highest possible degree of fineness of the raw materials, facilitating the reactions that take place during sintering; atomisation aims to remove the water from the barbotine from the grinding cycle; pressing aims to shape the ceramic mass through high-pressure compaction; drying aims to remove the water from the forming mass; and finally sintering of the formed mass. Below is a more detailed description of each stage mentioned above.

3.4.1.1 PARTICLE SIZE CONTROL

The greater the degree of fineness of the raw material, the greater its specific surface and consequently the greater its reactivity in sintering, i.e. the reactions that take place during sintering are anticipated. The milling residue from porcelain tile production must be between 0.5 and 1% when passed through a 44 p.m. mesh, which corresponds to average particle diameters of around 0.02 mm. This fineness has a major influence on the densification and vitrification reactions of the final product (ALBONETI, 1995).

3.4.1. 2DOSAGE

The raw materials can be dosed by traditional mechanical means or by deformable load cell systems fitted to the feed bins. This type of dosing is suitable for discontinuous milling processes. Dosing by this type of process consists of removing the raw materials from the storage bins using mechanical shovels,

unloading them into the dosing bins, dosing the raw materials with weighing belts and finally unloading the raw materials into the mill. In the case of continuous milling, a weighing and dosing system is used for the raw materials controlled by a computerised system, intermediate storage of the dosed ceramic mass in a silo that will feed the mill continuously during the milling process. This process also involves the prior mixing of the raw materials with the deflocculant and the aqueous suspension containing the screening residues, which are generally subjected to recirculation (SAINZ at al., 1999).

3.4.1.3 MILLING

Grinding is a critical stage, where control over particle size must be maintained to guarantee the compaction conditions and characteristics of the sintered product. Raw materials can be ground in the **presence of water or not, and are classified as "wet" or "dry"**, respectively. The main aim of reducing grain size is to increase the surface area of the raw material and, consequently, the reactivity between the materials. Wet milling gives the best results and is therefore used in the production of porcelain tiles. This process consists of wet milling the raw material and drying the barbotine in an atomiser. The greatest benefit of this process is that it homogenises a wide range of raw materials, leaving them extremely fine and consequently obtaining very fluid powders that ensure the ideal filling of the mould. At the end of the milling process, there is an aqueous suspension of the finely ground raw materials, the barbotine, with a water content that varies depending on the type of material. This content is around 30 to 40 per cent. The atomiser is used to remove this water (SAINZ at al., 1999).

Grinding can be done in continuous or discontinuous mills. In the latter, the grinding time for a 34,000 litre mill, for a monocoloured mass, is 18 hours to obtain a barbotine with residues ranging from 1.0 to 1.2% on a 44 pm mesh sieve. For coloured ceramics, the time varies from 14 to 16 hours for a barbotine with residues ranging from 2.0 to 2.5%, also on a 44 pm mesh. During the milling process, the viscosity, density (weight/litre) and residue (44 p,m mesh) must be checked to ensure good performance from the atomiser. If the barbotine has a very high viscosity, in the range of 2.0 to 2.2 °E, it must be corrected by adding water or a fluidiser so that the discharge time is in the range of 40 to 50 minutes (BIFFI, 2002).

3.4.1.4   ATOMISATION

It consists of dehumidifying the barbotine obtained at the end of the milling process, as mentioned above. The water content is controlled **using an atomiser, also known as a "spray dryer". The process consists of injecting the** conveniently nebulised **barbotine at high pressure (22 atm)** into a drying chamber, where it comes into contact with air at a temperature of between 500 and 600°C. With this sudden change in temperature, the water evaporates almost instantaneously, due to the high heat exchange coefficient caused by the accelerated movement of the particles, the high specific surface area of the droplets and the high temperature gradient between the air and the barbotine. At the end of this process, it is possible to obtain rounded grains with moisture and particle size distribution suitable for pressing (BIFFI, 2002). When atomising, the viscosity range of the barbotine must be kept very close in order to guarantee the stability of the atomised pattern. Variations in grain size mean variations in the colour of the finished product. The resting time of the freshly atomised mass must not be less than 36 hours, in order to guarantee homogenisation of the humidity.

3.4.1.5GRANULOMETRY OF   THE PRESSED CERAMIC MASS

Based on the characteristics of the granules that make up the ceramic mass for pressing, the pore structure of the ceramic compact is defined. Knowing that the granules are nothing more than agglomerates of primary particles joined together by secondary bonds, it is expected that they will have a certain amount of pores in their volume, which is known as intragranular porosity. It is also known that when the granules fill the mould cavities, voids will always form between them, no matter how efficiently they are packed. This second set of interstices formed during the packing of the granules is called intergranular porosity. The set formed by the intragranular and intergranular pores during the filling of the mould cavities defines the initial arrangement of pores in the ceramic compact, Figure 5 (ARANTES, 2001).

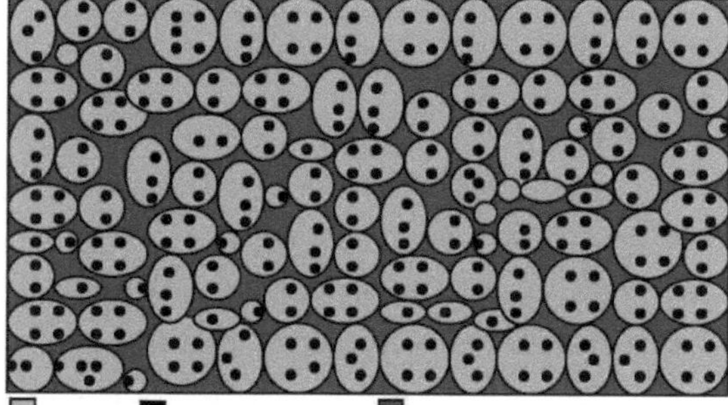

Figure 5 - Initial arrangement of intragranular and intergranular pores after filling the mould.

It is important to emphasise that there is a difference in size between intragranular and intergranular pores, with intergranular pores being considerably larger than intragranular pores. The total volume of these intragranular pores depends mainly on the shape and distribution of the particles that make up the granule. An adequate particle size distribution promotes the packing of particles with a high relative density, minimising the volume of these pores in the compact (ALBERO, 2000). In the case of atomised products, which represent practically all current porcelain tile production, a high solids content in the barbotine is also important, without, however, jeopardising the effectiveness of the atomisation process (RIBEIRO, 1999). In the case of the total volume of intergranular pores, this depends on the size distribution of the granules, the shape of the granules, their surface texture and the fluidity of the powder. Maximum packing is achieved with spherical granules, free of craters or hollows, with smooth surfaces and an adequate size distribution. The granules obtained must have low mechanical resistance so that they are broken during the pressing stage, thus creating a microstructure with less porosity (HECK, 1996).

3.4.1.6   CONFORMATION

During the porcelain tile production process, the forming phase is responsible for leaving the ceramic body with the highest possible green density, but compatible with the gas release problems that occur during sintering, the phase where the maximum densification of ceramic materials occurs. In current processes, the forming pressure used is around 45 MPa. This pressure generates a ceramic body with a density ranging from 1.38 to 1.96 g/cm^3 , which represents around 75% of the final density of the sintered product. This value has shown good results in the product after sintering (ALBONETI, 1995).

Pressing is the process used to form porcelain tiles and ceramic floor and wall tiles. In this forming process, the powders have moisture contents that generally vary between 5.0 and 5.5 % and are compressed between two surfaces, one movable (punch) and the other fixed (stamp). The pressure applied causes a reduction in the volume of both the intergranular and intragranular porosity, resulting in a strong densification of the powder through the rearrangement and partial deformation of the granules, allowing for a high green density of the ceramic body.

This reduction in pore volume is achieved through three different mechanisms: the displacement and rearrangement of the granules; the plastic deformation of the granules, the first two of which are related to the reduction of intergranular porosity; and the deformation and rearrangement of the primary particles that make up the granules in order to achieve greater packing, which is related to the reduction of intragranular porosity. It should be noted that almost all of the pore volume eliminated during compaction corresponds to intergranular pores (ALBERO, 2000; REED, 2000).

The greater the degree of compaction of the powder, i.e. the greater the green density, the greater the contact surface between the granules, thus increasing the possibility of reaction between them during sintering (ALBERO, 2000). The high forming pressure influences the absorption and shrinkage values of the sintered material. In the case of very fine powders, the forming pressure can reach 50 MPa. This value is generally used for single-colour products, as they have a finer grain size due to a longer milling time. Decorated products are critical because the deviation in the planarity of the pieces must be minimal, since in the polishing process the layer removed must be uniform to avoid differences in tone. When making products decorated with soluble salts, the temperature of the piece must be strictly controlled in order to guarantee a penetration of 1.5 to 2.0 mm. Moisture control during pressing is a must, as defects such as edge swelling can occur when the moisture content is less than 4% (BIFFI, 2002).

3.4.1.   7DRYING

The purpose of the drying phase is to reduce the amount of water used for moulding so that the material can be sintered in an industrially acceptable time and, in the case of mono-firing, to increase the mechanical strength of the green support, with a breaking load value of between 17 and 20 kg/cm^2 , so that it can be transported on the glazing lines. Due to its relatively non-plastic structure, porcelain tiles do not present any particularities or substantial differences in the dryer's cycle or working conditions. At this stage of the process, care must be taken with the drying speed of the tiles, as drying the surface too quickly can lead to cracks appearing when the internal water tries to migrate out of the tile. The dimensional variations caused by drying must also be controlled as a function of the amount of water released, so that there are no differentiated shrinkage zones that induce the appearance of internal stresses in the material (BIFFI, 2002; VIEIRA, 2003).

3.4.1. 8SINTERISATION

In the sintering process, two opposing phenomena occur concomitantly during part of the cycle. The first mechanism, the formation of a glassy phase responsible for densifying the body, occurs due to the use of a high content of fluxing raw materials in the formulation of the mass. This highly viscous glassy phase formed by capillary force establishes a laminar flow that gradually reduces the volume of the pores, promoting densification of the ceramic body.

At the same time, during the thermal cycle, the gas inside the compacted ceramic body, dispersed throughout its volume, and any gases formed as a result of the oxidation reactions of organic material and the decomposition of minerals in the mass, must be eliminated,

transported to the external environment as the sintering process progresses. However, from a certain stage in the sintering process, i.e. when the apparent porosity of the ceramic body tends to be zero, the glass phase formed involves practically all the pores. Thus, the gases that still exist inside the body are isolated from the external environment, giving rise to the closed porosity of the product (ARANTES, 2001).

Due to the high surface tension of the glass phase film surrounding the pores, the gas becomes trapped inside the compact, preventing it from escaping. From this point onwards, where the interconnection points between the pores and the external environment no longer exist, the two mechanisms come into direct contact. The gas trapped inside the pores makes it difficult for the laminar flow of the glass phase to advance and, as this flow progresses, the pressure of the gas trapped inside the pore increases as the pore volume decreases, making densification increasingly difficult. The increase in temperature, following the thermal cycle, also contributes to an increase in the internal pressure of the gases, while at the same time reducing the surface tension of the glassy phase, until a point is reached where the internal pressure of the gases trapped inside the pores exceeds the value of the surface tension of the glassy phase, causing the pore volume to increase, generating the phenomenon known as pore swelling (BELTRÁN, 2000).

The aim of sintering porcelain tiles is to vitrify the ceramic mass to the point of achieving zero or almost zero water absorption, in addition to achieving dimensional stability within the cycle considered. Among the various factors that contribute to achieving these objectives are: the reactivity between the raw materials that make up the ceramic mass; the degree of grinding of the barbotine; forming pressure and sintering temperature and cycle. Currently, the firing cycle for porcelain tiles ranges from 45 to 90 minutes for small formats and for large formats (60x60 cm) and with a large thickness (12 mm), respectively. For standard products, firing cycles range from 50 to 55 minutes at temperatures of 1210 to 1220 °C. In the case of masses containing refractory elements such as zirconium silicate and alumina, products whose high degree of whiteness is desired, the sintering temperature can vary between 1230 °C and 1250 °C. The product should remain at the maximum temperature of the firing cycle for 5 to 7 minutes and, at the end of sintering, should have maximum linear shrinkage values of around 7 to 8% for granulated products and 8.5 to 9.0% for single-coloured products (BIFFI, 1997).

This stage is the end of the porcelain tile production process, where it is checked that the fineness of the powder, the characteristics of the granules and the green density are within the ideal values for sintering. The time and temperature parameters must be strictly controlled throughout the cycle so that, in addition to quality end products, there is no waste of energy or raw materials. It is not common for slabs to expand after sintering, but this phenomenon can occur. This is due to situations such as: insufficient homogenisation of the raw materials during milling, which leads to little glass phase formation and consequently a high porosity index; the addition of minerals containing alkaline earth oxides (Ca, Mg) with the intention of obtaining a eutectic reaction that generates a greater quantity of glass phase with low viscosity; and the use of (potassic) feldspars that are not very melting at temperatures below 1200 °C (BERTHER, 1993).

3.4.1.9   POLISHING

The aim of polishing is to obtain slabs with an extremely shiny appearance, which is a feature that is highly sought after by coating consumers, as in addition to the aesthetic effect, it also makes cleaning easier. This operation is one of the last stages of the production line. At this stage, successively finer abrasives are used to obtain a smooth, shiny surface. There are two parameters to consider for a perfect polishing operation: the number of revolutions per minute of the abrasive heads and the number of oscillations and oscillating abrasive sectors. The end result of this operation is ceramic tiles that are perfectly polished to mirror level and ready for the successive squaring and bevelling phases (BIFFI, 2002).

3.4.1.10   GRINDING AND BEVELLING

In this operation, the ceramic slabs are squared off, which is carried out using diamond rollers with a high removal capacity, which work on the opposing sides of the ceramic slabs held in place and guided by overlapping belts. Since edge squaring allows the production of ceramic tiles of a single size, the operation is sometimes carried out on all the ceramic tiles leaving the kiln, even those not intended for levigating. After grinding, the bevelling phase begins, which is carried out using mandrels inclined at 45 degrees, equipped with both diamond rollers and traditional abrasives. Once the squaring and bevelling operation is complete, the ceramic slabs, always on the same line, are washed, cleaned and dried before being stored and/or dispatched to the market. (BIFFI, 2002) To summarise the above, the flowchart of the stoneware/porcelain tile production process is shown in Figure 6.

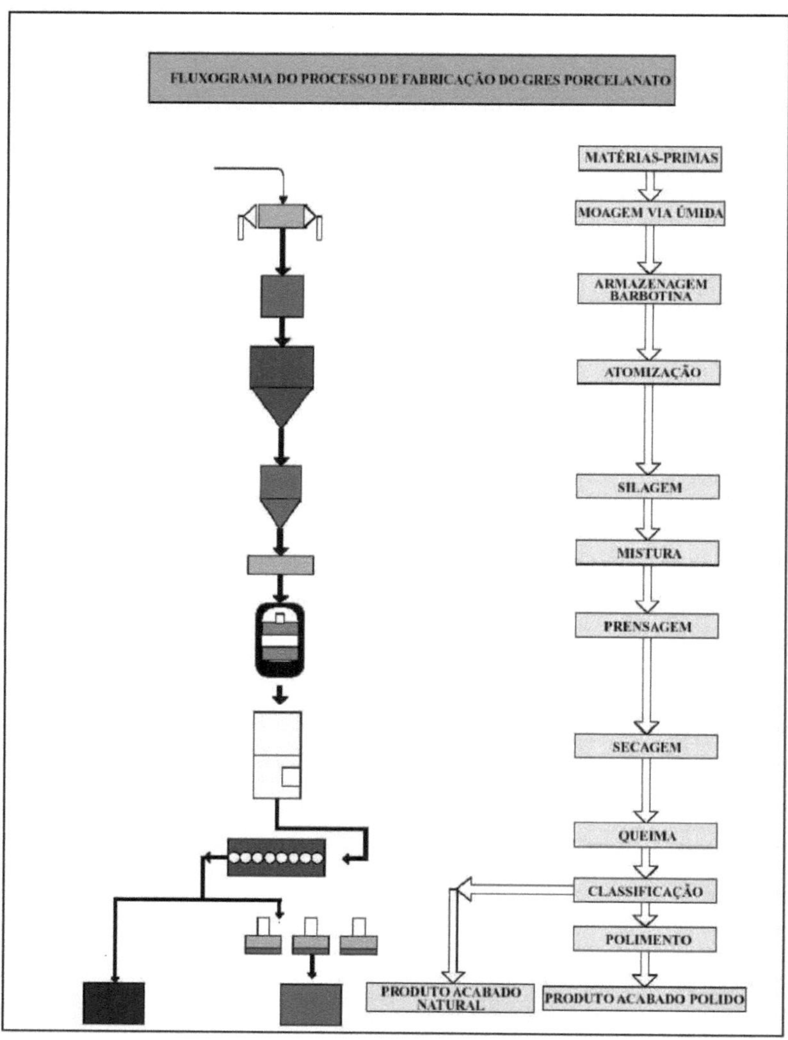

Figure 6 - Flowchart of the porcelain stoneware manufacturing process (ABCERAM, 2005).

3.4.   2STAINING PROBLEMS   DUE TO POLISHING

Despite being a product with low water absorption and consequently low apparent porosity, porcelain tiles typically have a certain volume of isolated pores inside the body, which constitutes the so-called closed porosity of the material, ranging from 5 to 12 per cent in various commercial products. During the grinding and polishing stage to which the pieces are subjected, a certain layer of around 0.5 to 1.0 mm of the product's thickness is removed, causing part of the pores previously isolated inside the body to be exposed on the surface, generating a series of irregularities, as shown in Figure 7 below (ARANTES, 2001).

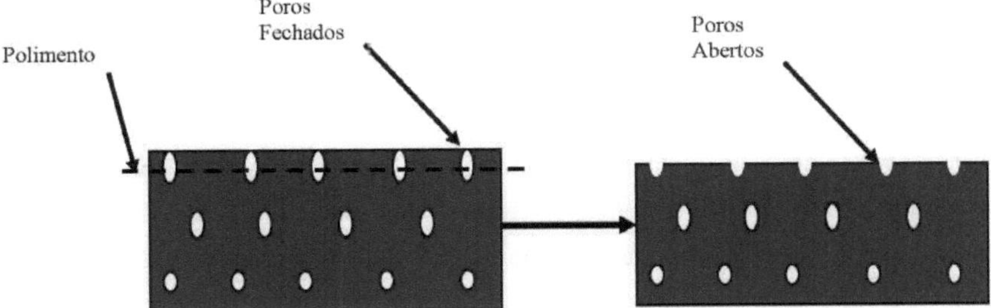

Figure 7 - Appearance of surface irregularities after the polishing stage.

The presence of irregularities on the surface of a ceramic body makes it easier for particles to adhere to its surface and more difficult to remove them. The resistance to staining of polished porcelain tiles is due to this fact. Greater or lesser susceptibility to the problem will be defined by the characteristics of these irregularities, which can be easily determined by the size, shape and texture of the pores present, as well as the number of irregularities (pores) per unit area (ARANTES, 2001).

In fact, studies indicate that there is a certain relationship between pore size and susceptibility to staining, with a value of around 30-60 p,m being the critical pore size at which the phenomenon of staining begins to occur. In view of this, it is important to obtain a minimum number of small, closed and isolated pores in the final microstructure of the product in order to maximise the part's resistance to staining. Closed porosity is an intrinsic characteristic of the product's densification process, and in general we can identify three stages of the production process that directly affect this phenomenon: the preparation of the mass, especially its particle size distribution; the forming process and the sintering stage.

The deformability of the granules is a characteristic closely linked to the porosity of the coating, since the more the granules deform, the more the interstices inside the piece are filled, reducing the value of the intergranular porosity of the body (BELTRÁN, 2000).

CHAPTER 4

WASTE USED IN THE CERAMICS INDUSTRY

The amount of waste generated in construction, the ceramics industry, the ornamental stone industry, the paper industry, sewage treatment plants and kaolin mining and processing shows a huge waste of reusable raw materials. The costs of this waste are shared by society as a whole, not only because of the increase in the cost of end products, but also because of the costs of removing and treating this waste which, in some cases, is disposed of illegally and inappropriately in riverbeds, on the banks of streets and roads in the suburbs and on vacant lots. However, the reuse of the aforementioned waste is practically restricted to its use as landfill material and, to a much lesser extent, for the conservation of dirt roads. This is due to a lack of awareness of their potential, as they should be seen as a source of highly useful raw materials for local industries. Educational institutions are therefore developing studies into various techniques that make it feasible to reuse waste from various industrial sectors and their reuse lines that guarantee the technical quality of the end products at lower costs than those produced in the conventional way (AMBIENTE BRASIL, 2005).

4.1   CONSTRUCTION WASTE

Recycling construction waste generates savings for entrepreneurs and benefits the environment. CONAMA Resolution 307 of 5 July 2002 establishes criteria, guidelines and procedures for the Management of Construction Waste, creating a chain of responsibilities that starts with the generator, passes through the transporter and reaches the municipalities. Each municipality is responsible for its waste, according to the resolution, and must establish actions to fulfil the established targets. Municipalities are prohibited from receiving construction and demolition waste in landfills, and must provide an area for depositing the specific material (CONAMA, 2002).

In Belo Horizonte, the waste resulting from construction work has a certain destination: recycling. The waste destined for recycling is made up of ceramic materials, concrete and mortar pieces, sand, gravel and so on. Since 1993, the capital of Minas Gerais has had a Programme for the Correction of Clandestine Deposits and the Recycling of Rubble, the aim of which is to promote the correction of the environmental problems generated by improper disposal. The recycled material is used in the construction industry to replace sand and gravel, or iron ore for road base and sub-base, according to the city's Urban Cleaning Superintendence's Social Communication Department. The city currently has 24 Small Volume Reception Units (URPVs), which receive up to 2 m^3 a day per carrier. These measures represent great gains for the environment: they reduce the need to create public areas to deposit this rubble; they also minimise the need to extract raw materials; and they reduce the deposit of materials in inappropriate places, which can result in a higher incidence of disease-transmitting animals. They also represent savings for the municipal administration, with fewer clandestine waste dumps and a reduction in diseases transmitted by the animals that live there. Contractors also benefit, as recycling waste results in savings on construction sites, without compromising quality (AMBIENTE BRASIL, 2005).

In Seropédica and Itaguaí, Rio de Janeiro, the extraction of sand for construction generates a large volume of waste called gum, which is deposited in pits or landfills and pollutes the environment. For this reason, the Federal University of Rio de Janeiro (UFRJ) has been working on recycling this material in the red ceramics industry. To this end, differential thermal analysis (DTA), thermogravimetric analysis (TG), X-ray diffraction (XRD) and X-ray fluorescence (XRF) tests were carried out. Five specimens with dimensions of 6 x 2 x 0.5 cm were formed[3] for each sintering temperature, 800, 900 and 1000 °C. The heating rate was 10 °C/min, with a 1-hour plateau for all temperatures. After sintering, the specimens were tested for water absorption (%), linear contraction (%), apparent density (%), loss on ignition (%) and flexural tensile strength (Mpa). The results obtained from the tests on the specimens showed that the residue can be used as a material for the ceramics industry, even at a temperature of 800 °C, which, compared to the other two firing temperatures, showed lower results in the mechanical tests. The temperature of 1000 °C is ideal when using this waste, with the following results: water absorption - 17.3%; apparent density - 1.84%; linear contraction - 3.3%; apparent porosity - 31.9%; loss on ignition - 17.1% and flexural tensile strength - 6.6 Mpa (MOTHÉ, 2005).

4.2   WASTE FROM THE CERAMICS INDUSTRY

Chamotte is solid waste from the manufacturing process of ceramic tiles, blocks and roof tiles, characterised by being a burnt product that has visual or dimensional defects that compromise its use. The volume of chamotte from the ceramics industry is generally disposed of in landfill sites. As for the materials produced in the firing process (tiles), the solid waste from the ceramic tiles produced is usually recycled by entering the process in the wet milling phase, forming the ceramic mass with other raw materials. This is possible because this coating is generally very thin (4.0 to 6.5 mm), fired at temperatures of up to 1090 °C. In addition, because they are more porous, they are more fragile. In the case of chamotte from gresified flooring manufactured using the single-firing process, recycling is not possible without proper prior processing, as the waste products are generally thicker than the double-fired products (6.5 to 11.0 mm), burnt at up to 1220 °C and denser. These products are therefore harder and more resistant to the comminution process normally used in the ceramic tile industry (CASAGRANDE, 2000).

At the Federal University of Santa Catarina (UFSC), a study is being carried out into the production of ceramic pavement mix using slag from gresified pavements manufactured using the mono-firing process, which has yielded good results. Laboratory studies have shown that it is perfectly possible to reuse ceramic tile slurry as a source of raw material to feed back into the production process itself at the grinding stage. From an economic point of view, this material can place ceramic manufacturers in a highly competitive position on the market. It also reduces the company's environmental liabilities while preserving the environment and people's health. Because some glazes contain lead; if the frit used is sodium-based, it solubilises in the presence of water and contaminates the soil; chamotte, after being turned into powder, is being used for other specific applications, currently under development by the group. The reuse process developed proved to be technically and economically viable for the reality of companies in the sector and the technology involved is compatible with current production demands and the layouts of ceramic factories, which further strengthens the idea of the viability of this procedure. This confirms that the basic principle of the process can be extended to the treatment of solid waste in other sectors such as construction, metallurgy (foundry sand) and thermoelectric plants (coal ash), among others (CASAGRANDE, 2000).

Also at the Federal University of Santa Catarina (UFSC), formulations are being developed for ceramic floor tiles made from solid waste generated in the Eliane Ceramic Tiles Group's ceramic tile manufacturing process. The raw materials (waste and plastic clay) and mixtures obtained, after appropriate processing, were characterised from a physical and chemical point of view and, at a later stage, compacted to obtain test specimens. The test specimens obtained, after a single firing cycle in a roller kiln, were characterised in terms of properties typical of finished ceramic products. The results showed that the masses considered met the requirements of the certification standards for finished ceramic products and are potential candidates for obtaining **"ecological"** ceramic pavements **with optimised properties and costs (BIFFI, 2002).**

At the Federal University of São Carlos - UFSCAR, the global ceramic properties of a clay residue from sand extraction in the Corumbataí region (SP) carried out by Sibelco were determined, with a view to its use in the manufacture of typical ceramic tile masses. The calculations developed during the characterisation are in line with the characteristics observed for the clay residue. The low plasticity developed is a consequence of the high content of non-plastic minerals (mica and hematite), despite their small average particle size (< 2 mm). The red firing colour is a characteristic commonly found in clays with high hematite contents, which was confirmed in the rational analysis. X-ray diffraction of the material from the analysis of the fraction retained on a 325 mesh indicates the majority presence of quartz, hematite and mica, which is also confirmed by the similar amount of residue found in the clay and the amount of original quartz and hematite in it. It can be seen that the sintering of the material takes place from 1000 °C, where the changes in properties become more evident. From the temperature of 1150 °C there is a very abrupt change in the colour of the body, which is evidence of super firing. This can be corroborated by the sharp decrease in the material's shrinkage rate, absorption and apparent porosity. The mechanical strength only rose sharply at a firing temperature of 1000 °C. Using the DTA and TG curves of the clay residue, it was possible to identify an endothermic peak in the DTA at approximately 555 °C, characteristic of kaolinite dehydroxylation, and an exothermic peak at approximately 915 °C, characteristic of mulititisation. The presence of kaolinite was also evidenced by the TG curve, where there was a drop in mass at 555 °C due to the outflow of water. The dilatometric curve presented by the clay residue showed an increase in the shrinkage rate above 900 °C, indicating that at this temperature the development of an appreciable amount of liquid phase begins. Based on the results obtained, it can be concluded that, in this first stage of the study, it is feasible to use the clay residue in red ceramic coating applications (WENDER, 2001).

At UENF, a study was carried out on the effect of firing temperature on the technological and microstructural properties of a kaolinitic clay incorporated with up to 20% chamotte, pressed at 20MPa and sintered in a laboratory oven in a range from 500 to 1100 °C. The following technological properties were assessed: bulk density, linear shrinkage, water absorption and mechanical strength. The results indicated that there was no significant variation in the properties evaluated between 500 and 900 °C. At 1100 °C, there was a sudden variation in linear shrinkage and diametrical compression, a reduction in the water absorption of the compositions and an improvement in mechanical properties due to the vitrification process. The reduction, albeit small, in the dry density of the clay was a result of the good particle size distribution of the chamotte, which facilitated increased packing. It was also found that the incorporation of chamotte above 5% is detrimental to diametrical compression (VIEIRA, 2005).

4.3 WASTE   FROM THE ORNAMENTAL STONE INDUSTRY

Due to the rise of the ornamental stone sector in recent years, there has been increased concern about the environmental impact caused by both the extraction of this material and its processing. For this reason, SENAI - SP carried out research and technological applications on samples of residual mud from the manufacturing and finishing process of marble factories, using it as a raw material in the red ceramics industry. Characterisations were carried out on both the clay and the mud, with granulometric analysis, thermal analysis and chemical analysis by X-ray fluorescence. The results of the chemical analysis showed that the mud has the characteristics of a dolomitic limestone material, both because it has a 9.73% MgO content (4.3% to 10.5%) and because it has a MgO/CaO ratio of 0.23 (0.08 to ,0.25). Clay, on the other hand, is considered a melting material because it has high levels of melting oxides such as Fe_2O_3, Na_2O and K_2O.

Based on the results, mixtures containing 0, 8, 16, 24, 32 and 40% slurry by weight in the ceramic mass were made. The masses produced were sintered at 6 different temperatures ranging from 850 °C to 1100 °C, with a variation of 50 °C, a heating rate of 40 °C/h and a 2-hour stop. After sintering, the specimens were subjected to the technological tests required by the standards.

Another characteristic of the addition of slurry to the mix is the loss of plasticity, which increases as the material is added. From the 16% mixture onwards, the doughs are considered to be not very plastic, but still with a sufficient level for extrusion. The particle size distribution curve of the raw materials showed that the slurry is mostly made up of fine materials, i.e. in a particle size that is acceptable for the ceramic production process.

The differential thermal analysis curves of the raw materials showed two endothermic peaks in the case of the sludge, one at 750 °C and the other between 900 and 950 °C, which when compared with the chemical analysis shows the decomposition of dolomite and calcite respectively.

The results obtained in the technological tests required by the Standard showed that the slurry reduces mechanical strength and increases porosity, which consequently increases water absorption, which is not very desirable, but if it is added at a content of 16 per cent by weight of the clay, it does not compromise the performance of the material according to the Standards in force (MELLO, 2005).

4.4   GLASS WASTE

Glass is used by the automotive industry in vehicle windscreens and in construction in windows and security partitions. Laminated glass is made up of two panes of glass bonded together by a polymer to form a **"sandwich". Before being shaped, the** glass **sheets** need to be cut and polished to give them the right shape for their application. During this stage, a large amount of solid waste is generated, consisting mainly of glass powder. This waste represents a problem from an environmental point of view, as it accumulates in factory yards without having any application. Studies have been carried out with the aim of finding technological applications for this material. Some of these studies have focused on the possible application of this powder as a substitute for quartz and feldspar in the manufacture of white ceramic tiles, others in polymer-based composites (VALERA et al., 2000).

At UENF - Universidade Estadual Norte-Fluminense, the effects of adding various types of glass on water absorption, firing shrinkage and tensile strength in ceramic masses for the production of roof tiles are being studied. The glass comes from broken empties, windows and TV tubes, the latter with a high lead content. Content of 6, 13 and 20% of the three glasses were added to the ceramic mixture from a roof tile production line in Campos, RJ. After mixing, chemical analysis was carried out on the mixture using atomic absorption. It was observed that the presence of K_2O and Na_2O type fusing oxides is very low, which will make liquid phase sintering difficult. However, the glass in the TV tube is lead glass, rich in PbO, which will play the role of a flux. Three formulations were made with contents of 6, 13 and 20 per cent of each glass and a reference mass with no added glass. The test specimens were fired at three different temperatures 900 °C, 975 °C and 1050 °C followed by a slow firing cycle of 2 °C/min up to 600 °C, remaining at this temperature for 60 minutes, after which they were heated up to 1200 °C at a rate of 2 °C/min, remaining at this temperature for 120 minutes. The cooling rate was 10 °C/min to 600 °C and 2 °C/min to room temperature.

After sintering, the specimens were subjected to the standard analyses. The results of the water absorption, linear shrinkage and tensile strength analyses, respectively, showed that as the temperature increased and the amount of glass added increased, the linear shrinkage and tensile strength also increased, while the opposite was true for water absorption. The linear shrinkage and tensile strength curves are related to the level of sintering in the structure, which promotes an increase in these and a decrease in water absorption. In this way, it was concluded that glass waste can be added to the mass without any damage to its properties analysed here, as long as it is in powder form and its granulometry is as close as possible to the granulometry of the mass (GODINHO, 2005).

4.5   WASTE FROM BURNING SUGARCANE BAGASSE

At the University of São Paulo (Unesp), the effect of adding 10 and 20% by weight of sugarcane bagasse ash to the mass was characterised. The test specimens were made with dimensions of 60 x 20 x 0.5 mm under a force of 7 tonnes, and sintered at 800, 900, 1000 and 1100 °C with a heating rate of 10 °C/min and a 1-hour plateau. The technological tests carried out were: water absorption, linear shrinkage, apparent porosity, loss on ignition, apparent specific mass and flexural modulus of rupture.

The results showed that the specimens with added ash had a lower flexural tensile strength.

On the other hand, they had a higher apparent specific mass, lower linear shrinkage, lower loss of mass in fire, lower apparent porosity and lower water absorption. As sugarcane bagasse ash is composed of silica, predominantly quartz, detected by x-ray diffraction analysis, the sample showed some refractory characteristics and low mechanical strength. The sample with 20% ash had the highest mechanical flexural strength when burnt at 900 °C (1.09 MPa), with the other values being less than 1 MPa for the other temperatures.

It was concluded that the addition of sugarcane bagasse ash improves some of the properties of the ceramic material and can be incorporated into the ceramic mass. Controlling the size of the ash powder particles can result in pieces with better properties than those obtained in this work (TEIXEIRA, 2005).

At the Universidade do Norte Fluminense - UENF, in Campos dos Goytacazes-RJ, the influence of sugarcane bagasse ash granulometry on the firing properties of a kaolinitic plastic clay from the region was evaluated. Ash was used with a particle size of less than 20 (840 p.m) and 200 (75 p.m) mesh. Additions of 0, 5 and 30% by weight of ash were made to the clay. The specimens were prepared by uniaxial pressing at 20 MPa and sintered at 900, 1050 and 1200 °C. The technological properties of the specimens after sintering were: linear shrinkage, flexural tensile strength, bulk density, loss on ignition and water absorption. In general, what was observed was that there was no improvement in the technological properties of the clays, due to the large amount of quartz present in the ash, which acted as an inert material, slowing down the sintering reactions (BORLINI, 2005).

4.6   METALLURGICAL WASTE

The production of pig iron generates millions of tonnes of blast furnace slag every year, which has the potential to become a raw material for other industrial segments. The components of the blast furnace charge are basically the gangue of the metallic charge (ores), fluxes and coke. Blast furnace slag contains CaO, SiO_2, Al_2O_3 and MgO as the main constituents, and MnO, Fe_2O_3, and sometimes TiO_2 as secondary constituents. Studies have been carried out into the use of slag as a raw material for various products, one of which is the use of slag as an input in the production of glass suitable for the fibre industry. This has resulted in glass with more than 50 per cent slag in its composition, while maintaining the desired properties. With the advancement of the steelmaking process, more and more good quality pig iron and slag with a constant composition are being obtained, enabling safer routes for the reuse of this material (ROCHA et al., 1999).

At the Universidade do Norte Fluminense - UENF, the microstructural aspects and technological properties of red ceramics incorporated with steel slag were evaluated. The ceramic masses were prepared using a local clay and steel slag from the LD steel refining process, which was added to the clay in percentages of 0, 5, 10, 20 and 30 per cent by weight. Test specimens were prepared by uniaxial pressing at 20 MPa and sintered at 600, 850 and 1050 °C. The microstructure of the compositions was assessed by scanning electron microscopy and X-ray diffraction. X-ray diffraction of the slag showed the presence of crystalline phases predominantly formed by Ca and Fe. The phases rich in Ca were identified as carbonates and complex silicates. Fe is present in the form of magnetite (Fe_3O_4) and wustite (FeO). Low intensity MgO diffraction peaks were also identified in the form of periclase (free MgO). The chemical composition of the slag confirmed the XRD results. The particle size distribution of the slag was quite dispersed and coarse, with an average grain size of 7860 pim.

The dry density of the clay did not change with the addition of slag, but this was not the case for the bodies sintered at 650 and 850 °C, whose densities decreased when compared to the dry density, but this parameter did not change as a result of the percentage of slag added. The bodies sintered at 1050 °C had a higher dry density up to 10% slag, with a considerable drop as the percentage of slag increased. Analysis of the water absorption of the compositions evaluated showed that for the temperatures of 650 and 850 °C, the specimens did not show significant variations. The same did not occur for the temperature of 1050 °C up to an addition of 20%, which corresponds to the reduction of open pores in sintered parts. Above this value, the specimens show an increase in water absorption.

The behaviour of the firing linear shrinkage of the compositions studied showed that at temperatures of 650 and 850 °C, there was no significant variation in the linear shrinkage of the clay with the increase in slag. On the other hand, at 1050 °C, there was a decrease in linear shrinkage as a function of the percentage of slag added. This result can also be attributed to the loss of mass due to the decomposition of the calcium carbonate present in the clay.

In the case of the tensile strength of the samples, it was observed that it did not change for the dry samples with the addition of slag, but for the sintered material, there was an increase in strength, especially at 1050 °C. This increase is due to the sintering process, which provides strong consolidation between the particles, reducing porosity at high temperatures. With regard to the amount of steel slag in the composition, it can be seen that at 650 and 850 °C, there is no significant variation in the parameter up to 10%. As this percentage increases, there is a slight decrease in the tensile strength of the samples. At 1050 °C, there is an abrupt reduction in mechanical strength. This behaviour is attributed to two factors: the decomposition of calcite, generating porosity, and the inert behaviour of the slag.

After evaluating the results presented above, it was concluded that steel slag is a steelmaking residue with high levels of Ca, Fe, Mg and Si whose loss of mass at high temperatures is mainly associated with the decomposition of calcium carbonate. Its coarse particle size is unsuitable for direct use in red ceramics without prior sieving or comminution. The incorporation of steel slag into clay masses leads to significant changes in the technological properties of the final ceramic product. The results also indicate that the recycling of slag in red ceramics should be carried out at maximum temperatures of 850 °C, so that calcite decomposition does not occur, and for maximum quantities of 10%, so as not to cause an increase in porosity and consequently a decrease in strength (VIEIRA, 2005).

4.7 WATER AND SEWAGE TREATMENT PLANT WASTE - SLUDGE

At the University of Vale do Rio dos Sinos, RS, the possibility of recycling sludge from water and sewage treatment plants as a filler in the production of ceramic blocks was studied. The sludge was classified as Class II, non-inert, becoming inert after sintering. Geometric, physical and mechanical characterisation tests were carried out on the ceramic blocks. The sludge was used as a filler in two local clays used to produce ceramic blocks. Both the sludge and the clays were subjected to preliminary analyses in order to determine the best percentage of sludge to add to the mix. Liquidity limit and plasticity limit tests were carried out, obtaining 62.60 per cent and 46.25 per cent respectively. Liquidity limit and plasticity limit tests were also carried out, obtaining 52.90% and 34.80% for clay A and 46.2% and 26.2% for clay B, respectively.

After the analyses, the percentage of sludge was set at 2%, and the forming process was by extrusion. The test specimens were 6-hole ceramic blocks in commercial size, which were sintered at a temperature of 900 °C, with a 3-hour stop and a heating rate of 125 °C/min. The results of the water absorption and suction analyses were within the standard limits of 8 to 25%. The results of the compressive strength test were higher than the standard of 1 MPa. According to the results obtained, it was possible to recycle the sludge from the WTP, and all the values laid down by the standards were achieved (VELEDA, 2005).

The University of São Paulo - Unesp, evaluated the properties of ceramic masses incorporating WTP sludge, taking into account the dates on which the waste was collected. The sludge was collected every month for 12 months when the decanter was washed. Specimens with 0, 5, 10 and 20% sludge incorporated into the mass were formed, measuring 60 x 20 x 5 mm and weighing approximately 20g. After sintering at five different temperatures, 850, 900, 1000, 1100 and 1150 °C, they were subjected to the following technological tests: water absorption, apparent specific mass, loss on ignition, apparent porosity, mechanical flexural strength and linear shrinkage. The average percentages of sand, clay, silt and organic matter present in the sludge samples showed a percentage of clay that gives the ceramic mass a plasticity greater than the ideal value for making tiles and cladding. In general, it was observed that the addition of sludge worsened the properties of the ceramic bodies, as evidenced by linear shrinkage, which is below the limit values of 5%, up to a temperature of 1000 °C. At higher temperatures, linear shrinkage is greater than the limit value of 6%. The date of collection has no influence on this property.

Water absorption increases as the percentage of sludge incorporated increases. On the other hand, water absorption decreases with firing temperature and several samples showed water absorption of less than 25%. With regard to the month of collection, a small variation is observed, showing that sludge with more sand has higher water absorption.

The analyses of apparent specific mass and apparent porosity, respectively, worsen with the incorporation of sludge. The apparent specific mass, obtained at 850 °C, is higher than the established minimum limit (1.7 g/cm^3) for samples with 1% sludge, except for April and August. The other concentrations (15 and 20%) exceed the limit value from 1000 °C onwards, except for April, June and August. In these three months the concentration of sand in the sludge was higher than in the other months considered. The apparent porosity shows a behaviour consistent with that observed for the apparent specific mass, i.e. when the apparent specific mass increases, the apparent porosity decreases. At 850 °C, all the samples show apparent porosity greater than the maximum established limit value (< 35%). A sharp drop in apparent porosity is observed from 1000 to 1100 °C, especially in the samples with the most sand (April, June and August), probably associated with the formation of a liquid phase. This behaviour in apparent porosity and apparent specific mass may be associated with the higher concentration of non-plastic materials (sand and silt) in the sludge, which alters the behaviour of the specimens during drying and firing.

All the samples with the sludge mixture always have a mechanical flexural strength greater than 10 MPa and can be used to produce roof tiles (limit value > 6.5 MPa). The recommended value for solid bricks is 2 MPa and for hollow bricks it is 5.5 MPa. At 850 and 1000 °C, the incorporation of sludge worsened the mechanical strength of all the samples, even causing its value to decrease with the firing temperature. From 1000 to 1150 °C, all the flexural strength values increase again with temperature. This different behaviour of the mechanical strength at temperatures close to 1000 and 1100 °C, in addition to the non-plastic material of the sludge, may be associated with the mineralogical composition of the clays and the concentration of clay minerals, which undergo crystallisation and phase changes close to these temperatures.

Based on the results found, it can be concluded that incorporating sludge into the ceramic mass only improves the mechanical resistance property, while worsening all the others. Incorporating 10% sludge can be used for sintering at 850 °C, which does not significantly alter the properties of pure clay (ALÉSSIO, 2005).

CHAPTER 5

KAOLIN IN THE SERIDÓ REGION

In Equador, Rio Grande do Norte, kaolin mine workers risk going underground to escape unemployment and the lack of options to earn between R$200.00 and R$250.00 at the end of the month. After descending 20 metres on a precarious abseil with no safety equipment, you reach a network of tunnels that have been dug over the last two years. There are no support beams or shoring. Lighting is provided by candles, one, two or three in one hand, depending on the darkness of the underground path, and in the other hand, a pick and shovel. A 15 metre hole is not among the deepest. There are some that can be up to 70 metres deep. In these places, according to the prospectors, the candle has a dual function: to light the way and to check the amount of oxygen.

Every day, these workers haul around 10 tonnes of raw ore to be sold to companies that industrialise the product, under imminent danger, because in addition to falls on the way down, there are frequent tunnel collapses. Not only that, but shortness of breath is a constant problem among mine workers. Constant exposure to mine dust has led to the appearance of silicosis, a disease that affects those who are constantly exposed to very small solid particles. The continuous friction on the lung membrane causes the body to react by creating a lesion. The regeneration of this site leads to the formation of a more fibrous and less elastic tissue, different from the natural tissue of the lung. The constant use of simple, cheap masks could prevent inhalation of the material and future pulmonary fibrosis. These problems can extend to the local population if the waste from the kaolin mining and processing industries is not properly disposed of and treated (REVISTA AQUI VIP).

Waste from the kaolin mining and processing industries has great mineral potential, as it consists of kaolin, mica, muscovite and a small percentage of quartz. Minimising the negative environmental impact of indiscriminate disposal of this waste has prompted technological studies in this area. Currently, the main use of this waste is in the structural ceramics industry, i.e. in the manufacture of bricks and tiles for civil construction. In the Seridó region, including the municipalities of Junco do Seridó, Equador and Juazeirinho, there is intense artisanal kaolin mining activity in weathered pegmatite veins embedded mainly in the quartzites of the Equador Formation. The Run of Mine (ROM) produced in these primitive excavations is washed on site in small units, which are also technologically primitive and have low yields. With the kaolin washing residue, concentrates of muscovite mica, kaolin and quartz are obtained. The market importance of kaolin from the Seridó is due to its exceptional quality for coating and/or loading speciality papers, as well as applications in the cosmetics and hygiene paste industries.

Due to the great importance of kaolin, in the last two decades there has been a slow and gradual increase in market interest in industrial minerals to supply the ceramics industries installed in the Northeast and Southeast regions, which has kept the extraction of minerals from pegmatites going, especially in kaolin processing units, although they are considered rudimentary and are regionally known as decanters. These units each have an installed capacity of between 600 and 2,400 tonnes per year. There are also two other kaolin processing units in the region with an installed capacity of 12,000 tonnes/year. The products obtained are the result of washing and classification and have grain sizes below 100 and 200 *"mesh"*. *The kaolin produced is mainly destined for the paper, ceramics, paint, plastic and rubber markets. Bleaching tests carried out on kaolins from the region* have shown that it is technically feasible to increase the whiteness of kaolin by 7 points (REP. FED. BRASIL, 2003).

CHAPTER 6

APPLICATIONS OF WASTE FROM THE KAOLIN MINING AND BENEFICIATION INDUSTRY

Kaolin mining and processing waste is being widely studied because of the huge environmental impact it has when disposed of indiscriminately in nature. As mentioned above, the amount of waste generated by the kaolin processing industry in Rio Grande do Norte shows a huge waste of raw materials. The costs of this waste are distributed throughout society.

Thus, at the Federal University of Rio Grande do Norte (UFRN), in the ceramic materials laboratory, research has been carried out with the aim of developing ceramic floor tile formulations that meet the conditions laid down in Standard NBR 13818/1997 - Ceramic tiles for cladding - specification and test methods, in force, using waste from the kaolin processing industry in Rio Grande do Norte. The materials used in this research were kaolin waste and rejects from the structural ceramics industry. The ceramic mass was obtained by mixing the two residues in proportions ranging from 0% to 100% of both, always keeping the moisture content at 7% (±1) for forming the test specimens by uniaxial pressing in a hydraulic press.

The specimens were subjected to various sintering temperatures. After the sintering stage, the ceramic specimens were subjected to technological tests where, based on the results obtained, it was concluded that it is possible to generate products with good physical and mechanical properties using only a mixture of chamotte and waste from the kaolin processing industry. Chamotte, when present in greater quantities in the mix, appears to be a more suitable raw material for obtaining red-based coatings, with a sintering temperature of no more than 1050°C. Kaolin waste, on the other hand, when present in greater quantities in the mix, appears to be a more suitable raw material for obtaining light-based coatings with better physical and mechanical properties, withstanding sintering temperatures of 1250 °C. The practice of using waste as a substitute for natural raw materials makes it possible to achieve environmental and economic benefits and minimise negative environmental impacts (VARELA, 2005).

At the Federal University of Campina Grande - UFCG, a line of research is also being carried out using kaolin waste. In this case, kaolin waste is combined with granite waste. This mixture has been evaluated for its potential as a raw material in ceramic masses, for the production of roof tiles and ceramic blocks. Initially, the raw materials were processed and characterised, then four compositions were prepared and formed by uniaxial pressing into prismatic sheets at a pressure of 20 MPa. The specimens were sintered at temperatures of 800 °C, 900 °C and 1000 °C. After sintering, the specimens were tested for water absorption, linear shrinkage, apparent specific mass, loss on ignition and flexural tensile strength at three points. According to the results obtained, it was found that the values for the physical-mechanical properties of the specimens after sintering at 800 °C, 900 °C and 1000 °C are within the values specified by the standard and the tensile strength values are above 5 MPa, as proposed by Souza Santos, and these masses can be used for ceramic blocks and roof tiles. This shows the potential of using kaolin and granite waste in the structural ceramics industry (RAMALHO, 2005).

Another source of kaolin waste that is being explored is waste from the paper industry. Since 1999, Votorantin Celulose e Papel has been supplying the demand of four ceramics industries in Piracicaba, in the interior of São Paulo, which have been using around 900 tonnes of waste from this industry as an input in the manufacture of bricks. According to Votorantin Celulose e Papel, reusing the waste avoids the expense of digging ditches, maintaining the land and effluents and monitoring the water table. As a result, the company is managing to save R$33,500 a year, which considerably reduces the company's liabilities. The company's waste is made up of liquids and solids, approximately 40% of which is fibre and kaolin, which can be added to clay to formulate mass for the manufacture of ceramic blocks. One of the first companies to use the material began receiving 300 tonnes of waste per month - 120 tonnes of solid material and 180 tonnes of liquid. The first tests were carried out in 1995. Since then, the ceramics company has avoided extracting 300 tonnes of clay a month. According to the owner of the pottery, the company saves water and clay by using the waste (NOLASCO, 2005).

The effluent material accounts for 10 per cent of the composition of the ceramic blocks. Brioshi Ceramics produces around 800,000 bricks a month, earning around R$850,000 a year. The economic side is very important in a sector where ecological awareness has not yet reached its due importance. The studies carried out made it possible to identify different mass formulations made up of the waste and a local clay, which resulted in a percentage of between 10% and 30%, in order to obtain the best composition. During the tests carried out to determine the best composition, the hazardousness of the waste was also determined to see if the product could cause any damage to health. The results showed that the chemical composition of this mass is not harmful to human health. There was no change in the deformation of the ceramic body. The compound proved to be an excellent thermo-acoustic insulator and can be used in the production of blocks, bricks, sandwich panels, lining boards, among others (NOLASCO, 2005).

One segment in which kaolin processing residues can be used is in refractory mortars and concretes. Preliminary studies have shown that the residue, being finely divided and basically made up of kaolinite, is an excellent raw material for manufacturing a highly reactive metakaolinite

(BARATA, 1998), whose mechanisms of action in Portland cement-based systems are, in a way, similar to those of active silica. Both accelerate the hydration process of Portland cement because they are made up of extremely fine particles that act as nucleation points for the formation of calcium hydroxide, as well as reacting quickly with this hydration product.

The difference is that in metakaolinite, regardless of the content of incorporation, the pozzolanic reaction reaches its maximum period between 7 and 14 days, while in active silica, depending on the percentage, the effects of its pozzolanic reaction can be significant for later periods (WILD, 1995; 1996). In addition to the ease with which this waste can be recycled and managed, the production of a highly reactive metakaolinite for the construction market could be very interesting from an economic point of view for the companies that generate kaolin processing waste, given that the sale price of this type of pozzolan on the market is approximately R$ 1260.00 per tonne, while the price of the final product of these companies, kaolin coating for the paper industry, varies between R$ 390.00 and R$ 420.00 per tonne (BARATA, 2002).

CHAPTER 7

MATERIALS AND METHODS

This chapter presents a description of the raw materials, characterisation techniques, the respective parameters adopted to carry out the experimental part of this thesis and the field survey carried out to assess the real situation of kaolin mining and processing industries in Rio Grande do Norte, in terms of the disposal of waste generated during the production of marketable kaolin.

7.1   TECHNICAL VISIT

A technical visit was made to a kaolin exploration and processing unit, Figure 8, in the municipality of Equador-RN. Processing consists of wet disaggregation in equipment called a beater. The product of the disaggregation goes to the washing chutes, Figure 9, where the impurities are retained and the fine part is carried to the settling tanks.

Figure 8 - Kaolin processing industry.

Figure 9 - Washing chute.

The fine part goes to a settling tank where the dark impurities, manganite - manganese oxide, are removed (Figures 10 and 11).

Figure 10 - Settling tank I

Figure 11 - Settling tank II.

The overflow from this tank goes to other settling tanks, where the decanted material is removed manually using shovels, placed in cloths and taken to the press (Figures 12 and 13).

Figure 12 - Placing the kaolin in the hydraulic press to remove excess water from the kaolin.

Figure 13 - Hydraulic press for removing excess water from kaolin.

Once the excess water has been removed, the material is taken for drying, which can be done in one of two ways: in a drying yard, where it is placed on a kiln, or in another exposed to the sun, Figures 14 and 15 respectively. This product is intended for the ceramics industry.

Figure 14 - Drying kaolin on a patio over a kiln.

Figure 15 - Drying kaolin in an open-air courtyard.

The tailings from this process are placed on the sides of the roads (Figures 16 and 17), causing visual discomfort for passers-by and causing a negative environmental impact on nature and jeopardising the health of local residents.

Figure 16 - Tailings from kaolin processing in the Seridó region - Ecuador.

Figure 17 - Detail of the tailings from kaolin processing in the Seridó region - Ecuador.

Therefore, the study and development of techniques to enable this material to be used as a raw material for the production of porcelain tiles is justified for economic reasons, such as reducing the liabilities of the generating company and reducing the raw material costs of the ceramics industry that produces white ceramic tiles and porcelain tiles; and social reasons, such as minimising the costs of removing and properly disposing of the waste, minimising the negative environmental impacts due to the degradation of the fauna and flora of the deposit site and preserving the health of the local population.

7.2   RAW MATERIALS

Two raw materials were used: the residue from the kaolin processing industry in the municipality of Equador - RN, and the standard mass (MP) for the production of porcelain tiles from a ceramics industry that is already established in the market. The standard mass was already supplied dosed and atomised, exactly as it is used in the company's production line.

7.3   CHARACTERISATION TECHNIQUES

The raw materials were characterised through chemical analysis of the powder by X-ray fluorescence (XRF) and mineralogical analysis by X-ray diffraction (XRD). Based on the X-ray fluorescence and X-ray diffraction results, the rational analysis of the kaolin residue and the standard mass was carried out using the computer programme for the rational analysis of clay minerals "MIDS" developed at UFRN (VARELA, 2005). The X-ray diffraction results revealed that the residue is made up of kaolin, muscovite mica and a small percentage of quartz. The results of the analyses of the

44

standard mass provided the raw materials and their respective percentages used to prepare the mass used to make the porcelain tiles.

7.3.1 CHARACTERISATION EQUIPMENT AND PARAMETERS

To analyse the kaolin processing residue by X-ray diffraction, it was ground and passed through a 200 mesh sieve (75 pm). The standard mass was already prepared for analysis, as it was supplied after being passed through the company's atomiser. The equipment used was a Shimadzu XRD-6000 with a Cu tube ($\lambda = 1.54056$ A). The parameters used were a current of 30 mA, with a scan from 5° to 75° for 20, a speed of 1.5°/min, a step of 0.02° and a voltage of 40 kV. The phases of the raw materials were assessed by comparing the peaks generated in the diffractogram with the standard charts in the JCPDS system registered with the ICDD (International Centre for Diffraction Data). The X-ray fluorescence analyses were carried out using Shimadzu EDX-700 equipment. Raw materials with a grain size of less than 200 mesh (75 μm) and a vacuum atmosphere were used to carry out the analyses. The results obtained are in the form of the most stable oxides of the chemical elements present in the compositions of the phases that make up the raw materials. The limitation of the apparatus is between the elements Sodium (Na^{12}) and Uranium (U^{92}), i.e. only elements in this range are detected.

The particle sizes of the raw materials were classified using laser diffraction. The raw materials were analysed in a Cilas model 920L laser granulometer. The medium used for the analysis was a mixture of distilled water (10 ml) and neutral detergent (2 ml) for every 2 g of material. The raw materials were dispersed in the above-mentioned medium for 60 seconds using ultrasound and the result was provided by the "The *particle expert*" programme for this purpose. The kaolin residue was ground and sieved through a 200 mesh (75 μm) sieve to be analysed.

Differential thermal analysis (DTA), thermogravimetric analysis (TG) and dilatometry are techniques that assess the behaviour of a material as a function of the temperatures to which it is subjected during the heating process imposed. For the dilatometric analysis of the kaolin processing residue, a heating rate of 10 °C/min was used on a specimen compacted to 50 MPa until it reached its melting temperature. The DTA curves show that the material undergoes changes during heating or cooling. These energy variations can be the result of four main causes: phase changes, decompositions in the solid state, reactions with an active gas and entropy changes without enthalpy changes (2^a order transitions). With regard to TG, this method provides the mass variations that occur during heating of the material and which can have two causes, decomposition or oxidation (FIGUEIREDO, 1986). For a more accurate assessment when it comes to TG, the use of derivative curves is of fundamental importance, as this provides more detailed information on, for example, the temperature at which the greatest loss of mass occurs, or the possibility of separating two or more events occurring at very close temperatures. The TGA-51H Shimadzu thermogravimetric analyser and the DTA-50H Shimadzu thermodifferential analyser were used for the thermal analyses. The following conditions were used for both: particle size below 200 mesh (75 μm), mass of 15 mg each of the raw materials, synthetic air flow of 50 ml/min and heating rate of 10 °C/min between 27 °C and 1250 °C. The results were analysed and the curve obtained using Shimadzu's TA-60 programme for thermal analysis. Dilatometric analysis was carried out using BP Engenharia's RB-115 dilatometer, capable of reaching a maximum temperature of 1650°C; this has the RB-3000 dilatometric analysis software as an additional tool. A heating rate of 20°C/min was used until the temperatures at which the liquid phase began to form. The dimensions of the specimens were 8 mm in diameter and 5 mm in height.

7.3.2   PREPARING THE CERAMIC MASS

The kaolin residue was wet milled at a ratio of 40% water by mass in an eccentric ball mill, BP Engenharia - bottle volume 650 ml, for a period of one hour. After grinding, the barbotine was placed on a tray and dried at 110 °C for 24 hours in an oven, TECNAL - model TE-397/3. After the drying process, the ground residue was deagglomerated in a mortar and passed through a 200 mesh sieve and packed in plastic bags. The standard mass was supplied under mixing conditions.

Table 2 contains the formulations studied, obtained by combining different proportions of the processed kaolin residue and the standard mixture. These formulations were mixed and homogenised wet for one hour in an eccentric ball mill, the same one mentioned above. The resulting barbotine was then poured into a tray and dried at 110 °C for 24 hours in a TECNAL TE-397/3 oven. The different mixtures were deagglomerated in a mortar and packed in duly labelled plastic bags.

Table 2 - Formulations of the putties studied.

Raw materials	Composition (% by weight)						
	MP0	MP1	MP2	MP4	MP8	MP16	MP32
Kaolin residue (RC)	0	1	2	4	8	16	32
Mass Standard (MP)	100	99	98	96	92	84	68

7.3.3 FORMING THE SPECIMENS

The different formulations, in powder form, had their moisture content adjusted to 7% (±1) by mass and were then granulated through a 65 mesh sieve. The granulated powder was separated into 14 g portions to obtain test specimens (SPs) of the same mass. Five specimens measuring 60 x 20 x 5 mm were made for each composition by uniaxial pressing under a pressure of 50 MPa. The equipment used was a Schulz PHS 15t hydraulic press with a capacity of 15 tonnes.

7.3.4 DRYING AND SINTERING THE TEST SPECIMENS

After the pressing stage, the specimens were dried in an oven at 110 °C for 24 hours and the sintering process began in an electric furnace, JUNG - model 2314 with a maximum working temperature of 1400 °C, in groups of five samples for each sintering temperature. The compositions were heat-treated at 1210 °C, 1230 °C and 1250 °C, with the heating rate set at 20 °C/min until the final temperature was reached, and the time taken to plateau was set at 15 minutes (Figure 18), at the UFRN Ceramic Materials Engineering Laboratory (LMC). A group of samples representative of all the formulations, MP0 to MP16, was also taken to be sintered in the kiln of the ceramics industry in question. This procedure was adopted with the intention of verifying whether the ceramic masses obtained would be applicable in the industrial process, as well as comparing and establishing correlations with laboratory data.

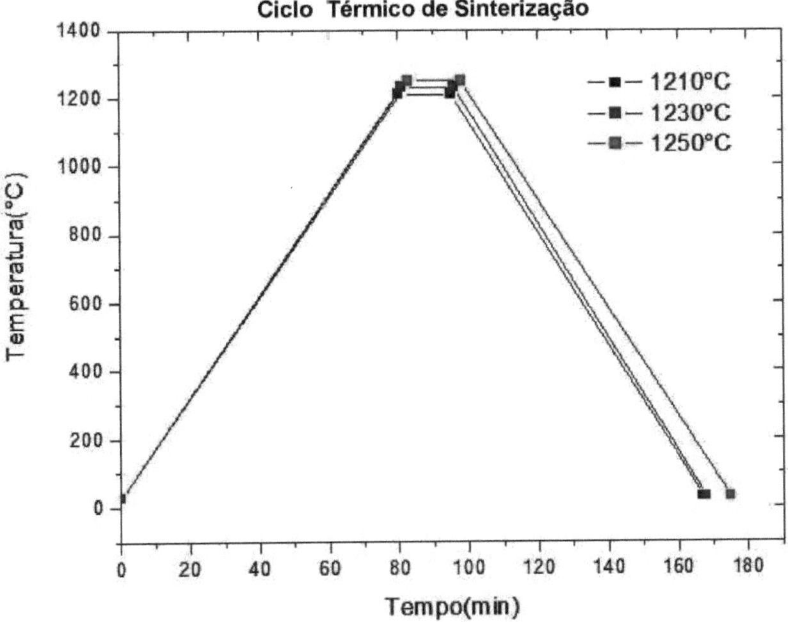

Figure 18 - Sintering thermal cycle at 1210 °C, 1230 °C and 1250 °C.

7.4 CHARACTERISATION OF THE FINAL PRODUCT

After the sintering process, the specimens were subjected to technological tests to determine water absorption (**AA**), apparent porosity (**PA**), linear shrinkage (**RL**), apparent specific mass (**MEA**) and three-point flexural modulus of rupture (**MRF**), according to the criteria set out in NBR

13818/1997 - Ceramic tiles for cladding - specification and test methods, briefly described below, and X-ray diffraction (XRD) and scanning electron microscopy (SEM) analyses to analyse the evolution of the ceramic tiles.

microstructure and fracture surface, as well as the presence of liquid phase formation and porosity.

7.4.1   WATER ABSORPTION   (AA)

Water absorption is the percentage of the mass of water absorbed by the body after sintering. The water absorption test was carried out as follows: the ceramic bodies were weighed immediately after leaving the kiln on a TECNAL - Mark 4100 analytical balance with a capacity of 4100g and a precision of 0.01g. They were then submerged in distilled water for 24 hours in a glass container. After this time, they were removed from the container and the excess surface water removed with a damp cloth and immediately weighed to check the variation in their new masses. The water absorption value, in mass percentage, was obtained using Equation 1 below.

$$AA(\%) = \frac{(M_U - M_Q)}{M_Q} x100;$$

(1)

Where AA is the water absorption in percentage; M_u is the mass of the specimen saturated in water; Mq is the mass of the dry specimen. After calculating the water absorption of each specimen, the arithmetic mean of the values obtained for each group was taken.

7.4.2   APPARENT POROSITY   (PA)

Calculating the apparent porosity gives the probable percentage of the volume of open pores after sintering the specimens in relation to their total volume. This value was calculated as follows: after weighing the specimens to calculate water absorption, the mass of the immersed specimens was also measured using the hydrostatic balance method. With the three values, M_u, **Mq** and the mass of the immersed specimen (M_i), Equation 2 was used to obtain the apparent porosity value as a percentage.

$$PA(\%) = \frac{(M_U - M_Q)}{M_U - M_I} x100;$$

(2)

Where **PA** is the calculated value of the apparent porosity and M_i is the mass of the specimen immersed in water.

7.3.3   LINEAR SHRINKAGE AFTER FIRING (RL)

Linear shrinkage considers the variation in the linear dimension of the ceramic body, in percentage, after the sintering stage. A variation with a positive value characterises a shrinkage of the dimension initially considered, otherwise the ceramic body is considered to have expanded. The procedure adopted was to measure the lengths of the test specimens before and after sintering using a Starret 721 digital caliper, accurate to 0.01 mm. Equation 3 below was used to calculate the linear shrinkage values after firing.

$$RL(\%) = \frac{(L_O - L_F)}{L_F} x100;$$

(3)

Where **RL** is the value of the linear shrinkage, in percentage, of the specimen after sintering; **L0** is the value of the length of the specimen before sintering and **Lf** is the value of the specimen after the sintering process.

7.3.4 APPARENT SPECIFIC MASS (MEA)

The apparent specific mass (ASM) is the ratio between the mass of the dry specimen and its volume, or the ratio between the dry weight and the difference between the saturated weight and the immersed weight, as shown in Equation 4 below:

$$MEA(g/cm^3) = \frac{M_Q}{M_U - M_I};$$

(4)

The arithmetic mean of the groups of values obtained for each formulation was taken.

7.3.5 3-POINT BENDING TENSILE STRENGTH (TRF)

Flexural tensile strength refers to the material's resistance to simple bending using the three-point method, according to the method proposed by VICAT. To measure this property, a Shimadzu AG - I 250KN universal testing machine was used, operating at a speed of 0.5 mm/min, according to the method proposed by the ASTM D-790/1986 standard. To obtain the results, tests were carried out on five specimens for each of the formulations, with the final value given as the arithmetic mean of these values. The calculations were carried out automatically using Shimadzu's Trapezium 1.14 software. Equation 5 below provides the results.

$$TRF(kgf/cm^2) = \frac{3PL}{bh^2};$$

(5)

Where TRF is the tensile strength (kgf/cm^2); P is the load reached at the moment of rupture (kgf); L is the distance between the supports of the specimen; b is the width of the specimen and h is the height of the specimen.

CHAPTER 8

RESULTS AND DISCUSSIONS

This chapter presents the results of the tests and analyses carried out on the raw materials used in this thesis and on the final products obtained from sintering, at the three reference temperatures, the different masses produced with the introduction of kaolin processing residue contents ranging from 0 % to 32 %. This chapter also includes considerations of the results achieved in these tests and analyses.

8.1   CHEMICAL AND MINERALOGICAL ANALYSES

The chemical and mineralogical analyses of the standard mass, supplied by the ceramics industry, will not be disclosed in this thesis as this is confidential company information. Table 3 contains the results of the chemical analysis of the kaolin processing residue studied, in mass percentage, in the form of oxides. Table 4 contains the results of the X-ray diffraction analysis of the kaolin residue in question.

Table 3 - Chemical analysis of kaolin processing residue obtained by XRD.

Sample	SiO_2(%)	Al_2O_3(%)	Fe_2O_3(%)	K_2O(%)	CaO(%)	Na_2O(%)	MgO(%)
Kaolin waste	44,25	48,66	1,02	5,24	0,0	0,0	0,44

Table 4 - Qualitative analysis of crystalline phases in kaolin processing residue.

Sample	Present Phases		
Kaolin waste	Kaolinite	Mica Moscovita	Quartz

The results presented indicate that kaolin processing residue has a moderate alumina/silica ratio (>1:1) and a low iron oxide content, approximately 1%, and can behave as a structure-forming raw material even at high temperatures, above 1100 °C. Between 500 and 600 °C kaolinite loses its structural water and converts into metakaolinite, whose amorphous structure eliminates amorphous SiO_2. As heating continues up to around 950 °C, the formation of YAl_2O_3 spinel or Al-Si spinel begins, with a theoretical composition of $2Al_2O_3.3SiO_2$. The first liquid phase formation theoretically begins at 985 °C, the eutectic temperature of the feldspar-K + SiO_2 mixture. From there, the liquid formed reacts with SiO_2 eliminated from the metakaolinite, since the SiO_2 in the quartz will only be attacked at higher temperatures. The remaining Al_2O_3 and SiO_2 from the metakaolinite then form the primary mullite with very small lamellar crystals that appear in aggregates where kaolinite used to be. As the temperature rises, the formation of the liquid phase increases with the fusion of K-feldspar and muscovite mica at around 1150 °C. Thus, the primary mullite aggregates are penetrated by alkali ions from the feldspar and muscovite mica and the alkali depletion of the K_2O - Al_2O_3 - SiO_2 system allows the secondary mullite to crystallise. As the temperature rises to 1200 - 1230 °C, the dissolution of the secondary mullite begins as a result of the enrichment of the liquid phase in SiO_2, when the quartz begins to be attacked and dissolved, increasing the viscosity and consequently the toughness of the liquid phase.

Another point noted is the high potassium oxide content in the kaolin residue. The presence of this oxide should lead to the formation of an abundant liquid phase above 1100 °C. Due to the high alumina content and low Fe_2O_3 and MgO contents, this liquid phase can behave as a high viscosity liquid phase and/or as a transient liquid phase, which is desirable for obtaining products with higher physical and mechanical properties. The standard mass, in turn, has all its contents controlled by the production process adopted by the industry.

Figure 19 shows the X-ray diffractogram of the kaolin residue. The result shown indicates that the kaolin residue is made up of kaolinite, mica, muscovite and quartz, which is in line with the results obtained in the X-ray fluorescence analysis, where the oxides constituting these phases are shown and theoretically quantified by the rational analysis presented in Table 5 below.

Table 5 - Rational analysis of kaolin processing residue.

Samples	Quartz	Hematite	Mica Moscovita	Kaolinite	Accessories*
Kaolin waste	4.0	1.0	45.0	49.0	1.0

* All components that for some reason were not identified in one of the aforementioned analyses are included.

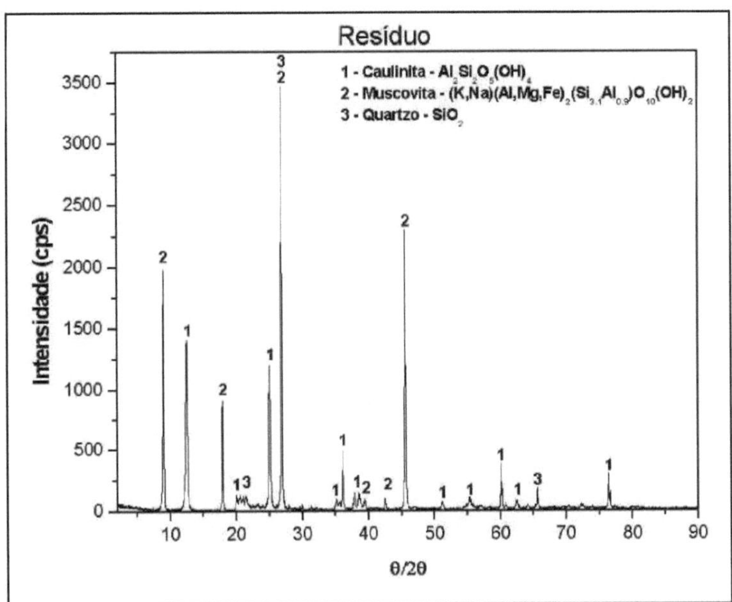

Figure 19 - X-ray diffraction analysis of kaolin residue.

Based on the results obtained from the analyses of the kaolin residue, it was found that it is made up of kaolinite and muscovite mica as the main phases and quartz as a minority phase. Kaolinite is an excellent structure-former over a wide range of firing temperatures and muscovite mica can act as a flux at higher sintering temperatures. Quartz reduces the shrinkage of the mass, since during the formation of the liquid phase it behaves as if it were the material's skeleton. Despite being present in small quantities, the behaviour of Fe2O3, hematite, in the mass produced must be observed, as iron acts as a flux at lower temperatures and at higher temperatures it can generate gas release reactions, causing excess dimensional variation, as well as forming bubbles in the sintered body. Another important point is the colouring effect caused by iron oxide when present in quantities greater than 2%, which does not occur in this case, as presented in the rational analysis and X-ray fluorescence above.

8.2 PARTICLE SIZE ANALYSIS

Figures 20 and 21 show the results of the analyses of the particle sizes of the standard mixture and the kaolin residue, respectively. It can be seen that the average D50 values are in the order of 39 ⌊ im for the kaolin residue and 10 p,m for the standard mix. The D50 value of the residue is above the recommended value for producing porcelain tiles, i.e. close to 20 pim. However, an important point is that the particle size distribution range of the residue is within the range observed for the standard mix, so the addition of kaolin residue to the standard mix should not significantly alter its distribution curve after homogenisation. As can be seen during the practical part of this thesis, the green density of MP0 was 1.73 g/cm³ and that of the MP8 formulation was 1.76 g/cm³ . This behaviour is directly linked to the new particle size distribution, which used to be bi-modal, MP0, and after the mixing and homogenisation processes became tri-modal for the MP8 mixture, contributing directly and effectively to better packing of the mass at the compaction stage.

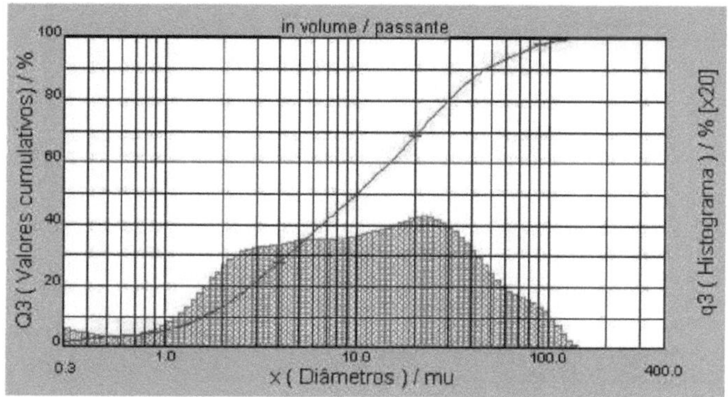

Figura 20 - Particle size analysis of the standard MP0, D50 10gm mixture.

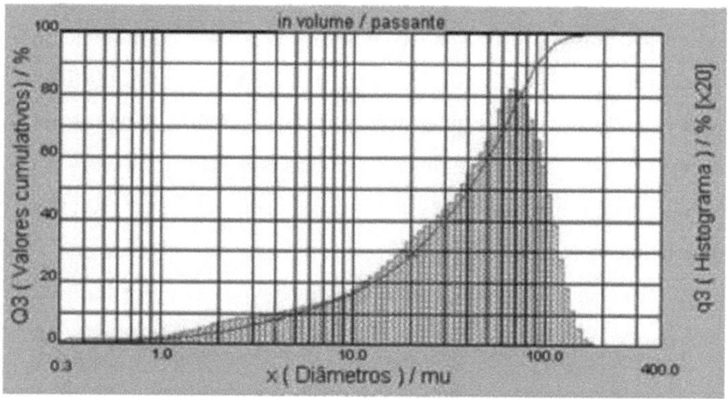

Figura 21 - Particle size analysis of kaolin residue, D50 from 39jim.

The fact that the distribution curve of the standard mass did not change can be seen in the results of the analyses shown in Figures 22, 23, 24, 25, 26 and 27, for the mass formulations MP1, MP2, MP4, MP8, MP16 and MP32, with 1, 2, 4, 8, 16 and 32% added kaolin residue, respectively. The average diameter of the resulting masses, D50, was 7.71 pm for the MP1 formulation, 8.01 pm for the MP2 formulation, 6.11 pm for the MP4 formulation, 6.31 pm for the MP8 formulation, 9.98 pm for the MP16 formulation and 11.26 pm for the MP32 formulation. The lowering of the average

54

diameter of the standard mass after the waste was added may have been due to the wet homogenisation process that was carried out in eccentric ball mills for 30 minutes. This contributes to an increase in the sintering power of the masses with the presence of kaolin residue. The only exception was the case of MP32, where there was an increase in the average diameter. This can be attributed to the need to increase the grinding time due to the large volume of residue added to the paste.

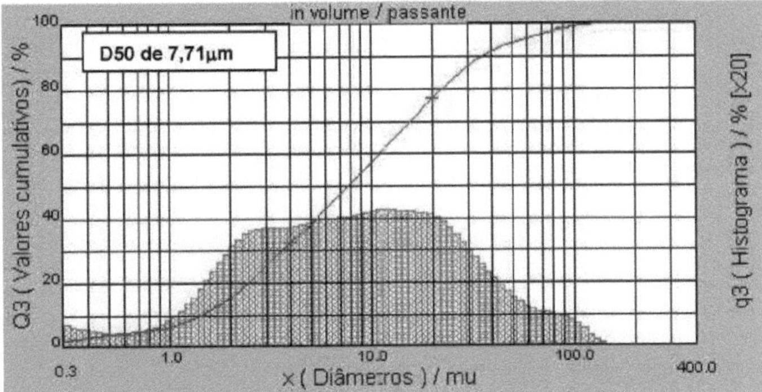

Figure 22 - Particle size analysis of formulation MP1 - MP + 1% kaolin residue.

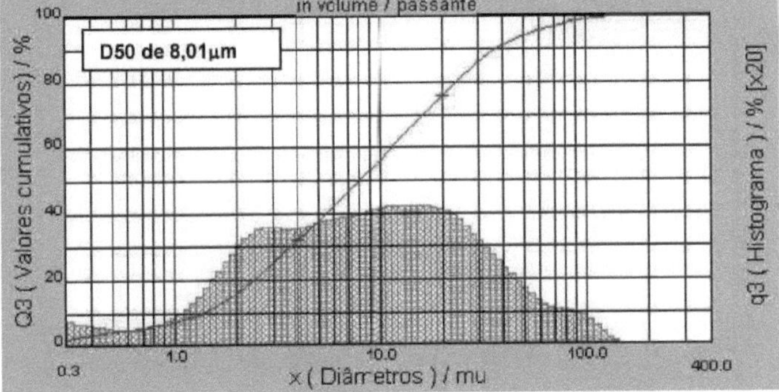

Figure 23 - Particle size analysis of formulation MP2 - MP + 2% kaolin residue.

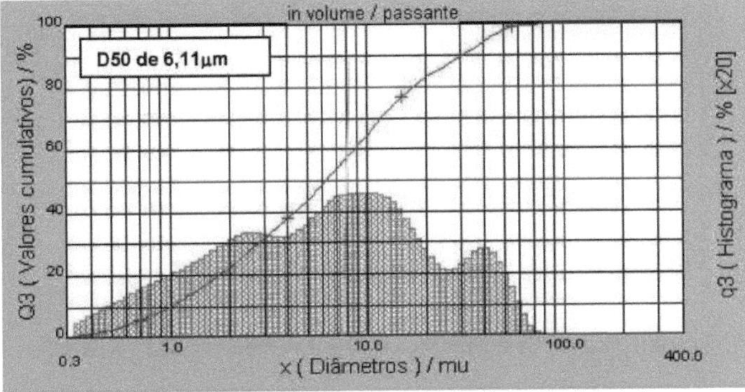

Figure 24 - Particle size analysis of the MP4 - MP + 4% kaolin residue formulation.

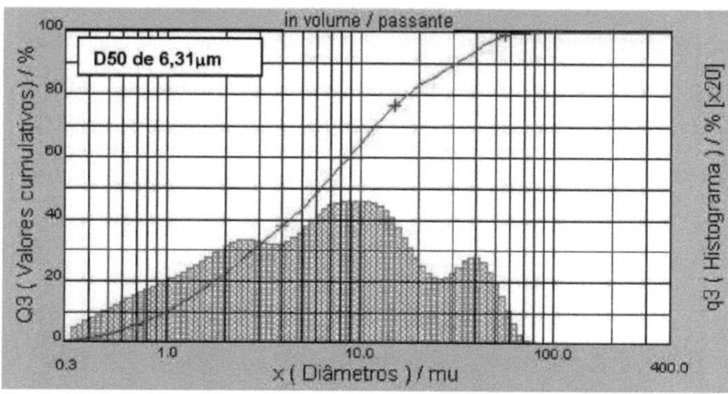
Figure 25 - Particle size analysis of formulation MP8 - MP + 8% kaolin residue.

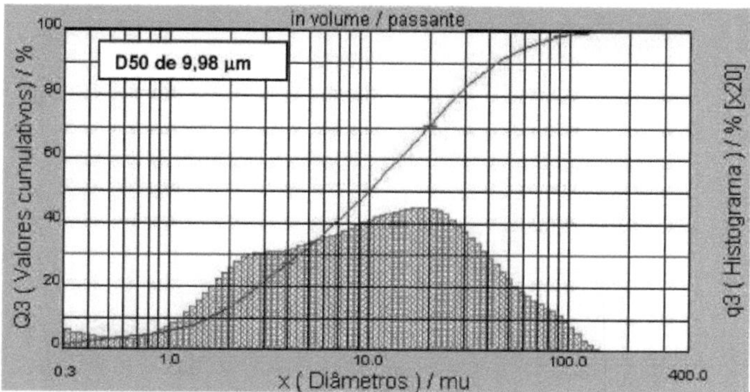
Figura 26   - Particle size analysis of the MP16 - MP + 16% kaolin residue formulation.

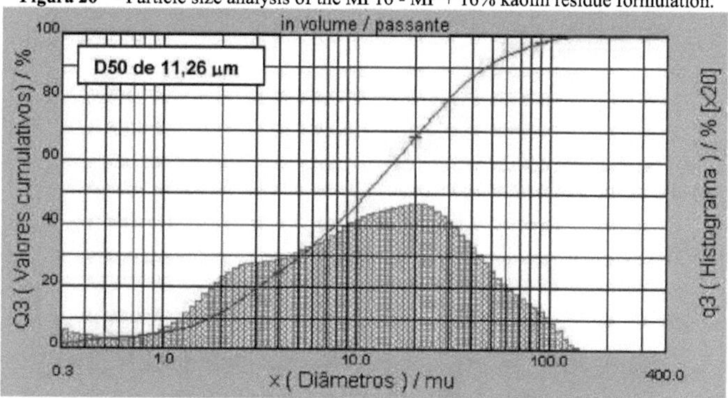
Figura 27   - Particle size analysis of the MP32 - MP + 32% kaolin residue formulation.

8.3   THERMAL ANALYSES

8.3.1   THERMOGRAVIMETRIC ANALYSIS AND DIFFERENTIAL THERMAL ANALYSIS

The thermogravimetric curve (TG) and the differential thermal analysis curve (DTA) of the kaolin residue, Figures 28 and 29, provide information on the mass loss and energy variations, respectively, that occurred during the heating of the sample, which serve as indicators of the possible reactions and transformations of the phases present in the material.

Thus, analysing the two curves in parallel provides a better understanding of the phenomena taking place. The DTA curve (Figure 29) shows an endothermic peak due to the loss of water from moisture, between the temperatures of 100 and 120 °C. At the same time, a mass loss of around 0.3% is observed in the thermo gravimetric curve.

Between the temperatures of 400 and 800 °C, the greatest mass loss occurred, 4.804%, due to the dehydroxylisation of kaolinite. The mass loss that occurred after this temperature range was attributed to the dehydroxylisation of muscovite mica, with a loss of 1.3% of the total mass loss, which was 6.4%.

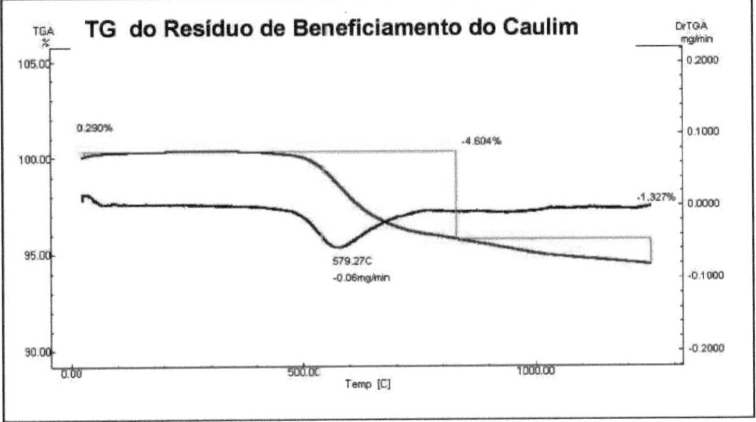

Figure 28 - Thermogravimetric analysis of kaolin residue.

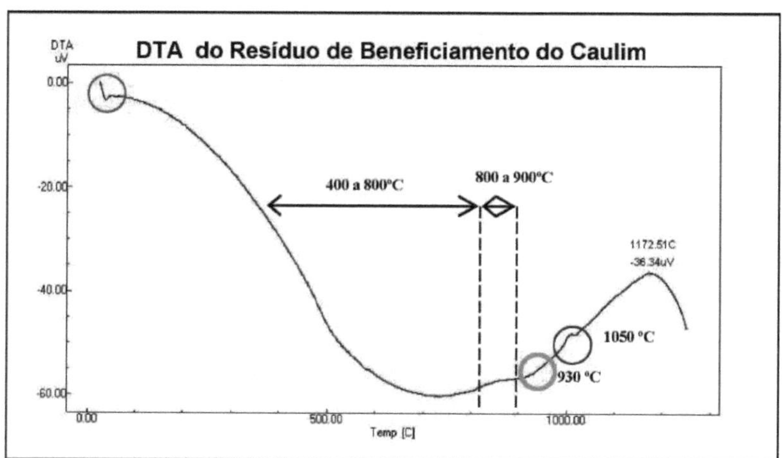

Figure 29 - Differential thermal analysis of kaolin residue.

In the temperature range between 800 and 900 °C, kaolinite is transformed into metakaolinite, as can be seen by the rise in the curve in this temperature range. Around 930 °C, the structure of metakaolinite is lost, a fact that is only present in well-crystallised kaolinites, which may represent the elimination of the remaining OH groups or the complete rupture of the SiO_4 group network, effects that may coincide, the former facilitated by the latter. From this point onwards, amorphous SiO_2 begins to develop and its maximum content is reached at 950 °C. From this temperature onwards, the amorphous silica content begins to decrease, reaching its minimum values around 1050 °C, where an exothermic peak is observed that can be attributed to the formation of the aluminium-silicon spinel or the nucleation of mullite. The exothermic peak, observed at around 1172.5 °C, can be attributed to the release of cristobalite when, from 1100 °C, the aluminium-silicon spinel is transformed into mullite, or to the formation of mullite from the spinel formed by muscovite mica, also present in the residue.

It is worth noting that the presence of MgO found in the X-ray fluorescence of the kaolin residue, as shown above in Table 3, and the CaO present in the X-ray fluorescence of the standard mass, which comes from calcium carbonate, can act as mineralisers or additives, accelerating the nucleation and growth of the mullite crystals, so two mechanisms can be considered:

$$\text{Metacaulinita} \xrightarrow[\text{CaO (Mineralizador)}]{\text{Sem segregação de } SiO_2 \text{ e } Al_2O_3} \text{Mulita}$$

$$\text{Metacaulinita} \xrightarrow[\text{MgO (Mineralizador)}]{\text{Com segregação de } SiO_2 \text{ e } Al_2O_3} SiO_2 + Al_2O_3 \rightarrow \text{Mulita}$$

8.3.2   DILATOMETRIC ANALYSIS

The dilatometric curve of the kaolin processing residue (Figure 30) shows its behaviour when heated at a rate of 20 °C/min. Up to a temperature of approximately 900 °C, the residue shows a linear expansion, explained simply by the linear coefficient of thermal expansion. In this temperature range, phenomena occur such as the elimination of moisture and adsorbed water in the form of gas, which occurs up to the maximum temperature of 200 °C, dehydroxylation reactions of the raw materials, $Mg(OH)_2$, found in X-ray fluorescence, and kaolinite. There is also the increase in volume suffered by quartz when it transforms from the *p-phase* to the p-phase, which begins at 573 °C and ends at 900 °C.

Between 900 °C and 980 °C, the specimen stops expanding and starts shrinking, thus showing a change in the slope of the dilatometric curve. At this point, vitrification begins with the release of cristobalite, which will react with free metal oxides to form glass, mainly alkaline, alkaline-earth and iron. From 980 °C to 1020 °C there is a slight expansion. From this temperature onwards, there is a sharp drop, which can be attributed to volumetric diffusion, which is responsible for shrinkage and densification, along with the formation of the liquid phase. The liquid coming from this region flows into the interstices between the more refractory particles, which have not fused, and by capillarity causes these particles to come closer together, resulting in quite significant shrinkage. The greater the amount of material that has melted and the lower the viscosity of the liquid formed, the greater the shrinkage.

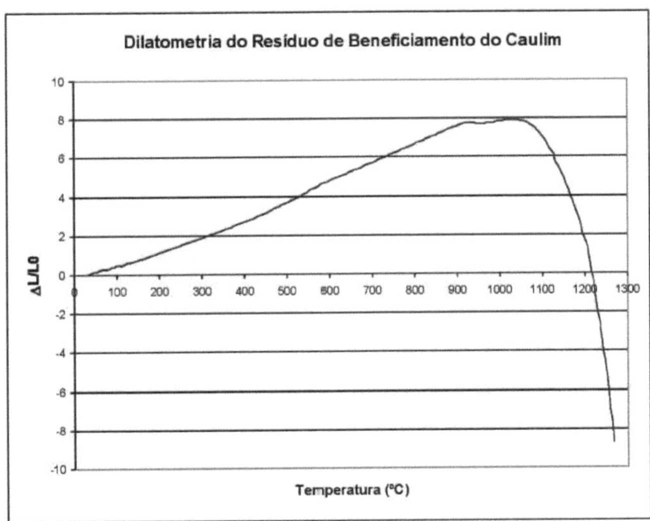

Figure 30 - Dilatometric analysis of kaolin residue.

In MP0, Figure 31, the same expansion behaviour can be seen as in the residue, up to a temperature of 930 °C. However, in addition to the facts mentioned above, there is also the presence of carbonaceous material, CaCO3 and the clays that are the raw materials that make up this mass. Thus, the loss of CO_2 from the limestone up to 930 °C and the organic matter and hydroxyls (OH) in the clays contribute effectively to the expansion shown in the dilatometric analysis of MP0 up to this temperature (Figure 31). Between 930 °C and 1020 °C, there is a slight shrinkage, which can be attributed to the rearrangement of the raw material and the start of vitrification with the release of cristobalite, which will react with the free metal oxides to form glass, mainly alkaline, alkaline-earth and iron. From 1020 °C to 1060 °C, there are no dimensional variations, but from this temperature onwards, there is a sharp drop in the dilatometric curve, also attributed to volumetric diffusion, which is responsible for shrinkage and densification, along with the formation of the liquid phase.

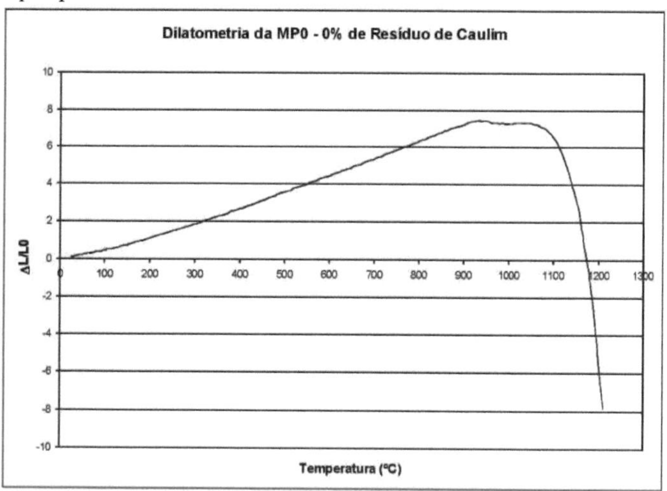

Figure 31 - Dilatometric analysis of MP0 - Standard Mass.

As can be seen in Figures 30 and 31, respectively, the kaolin processing residue shows a less steep downward slope in its dilatometric curve than the curve for the standard mass, MP0, which indicates a tendency for the liquid phase to form at a higher temperature than MP0.

Analysing the dilatometry of the new masses proposed, MP1 to MP16, shows that there is a positive interaction between the raw materials, since the downward slopes of the curves tend to become steeper, i.e. there is a tendency for the temperature at which the liquid phase begins to form in the proposed masses to decrease as the percentage of kaolin residue increases, as shown in Figures 32 to 36. This behaviour can be attributed to the existence of calcite and albite (rich in CaO and Na_2O) from the standard paste and kaolin residue (rich in K_2O - muscovite mica), highly melting oxides, which contribute to lowering the paste's melting temperature. In other words, the albite and calcite

60

would be potentiating the formation of the liquid phase by the muscovite mica. In other words, albite and calcite would be enhancing the formation of a liquid phase by the muscovite mica, thus promoting the anticipation of the sintering processes of the clay, contributing to the densification of the specimen at lower temperatures.

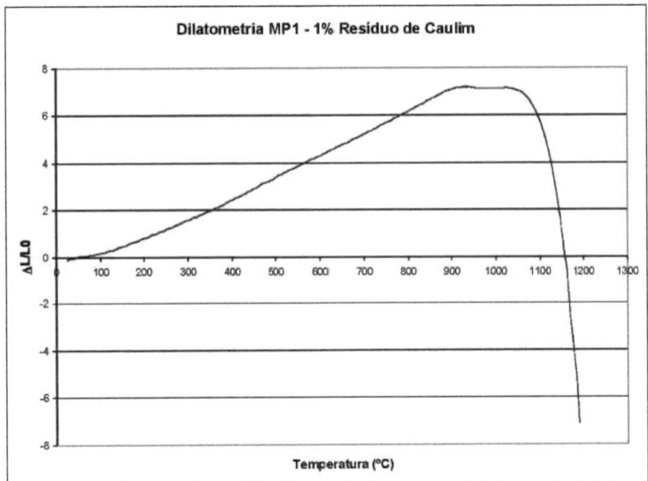

Figure 32 - Dilatometric analysis of MP1 - standard mass (MP0) + 1% kaolin residue.

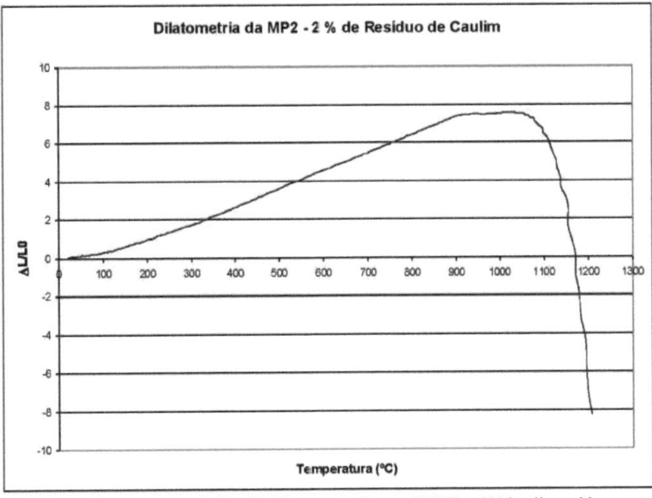

Figure 33 - Dilatometer analysis of MP2 - standard mass (MP0) + 2% kaolin residue.

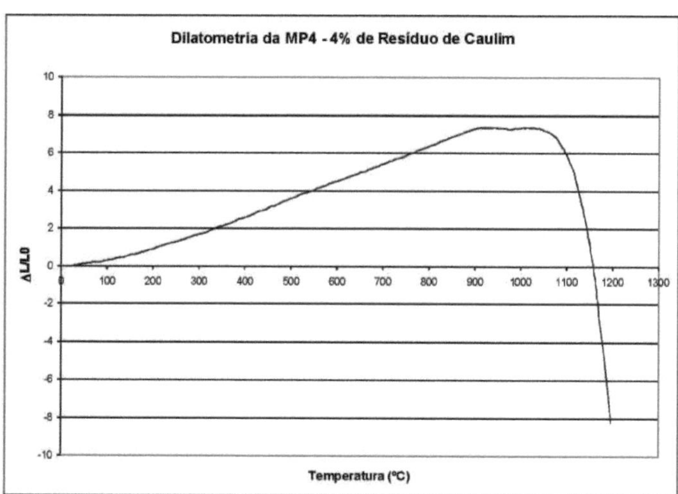

Figure 34 - Dilatometric analysis of MP4 - standard mass (MP0) + 4% kaolin residue.

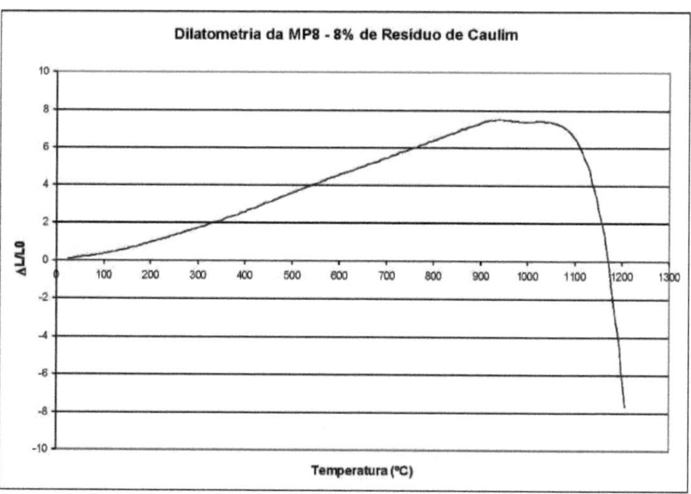

Figure 35 - Dilatometer analysis of MP8 - standard mass (MP0) + 8% kaolin residue.

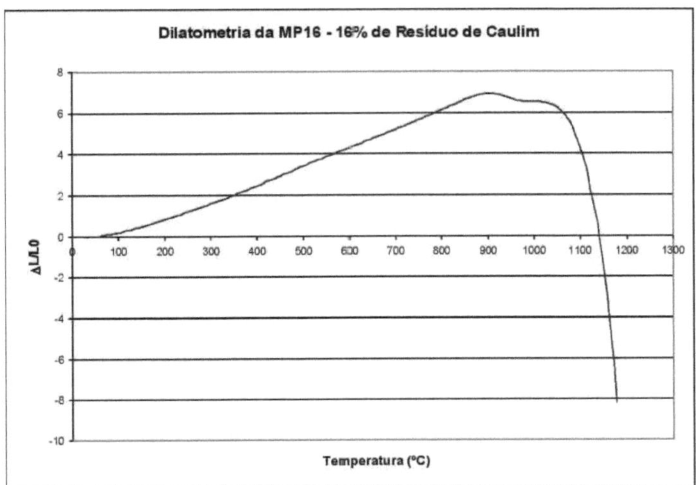

Figure 36 - Dilatometric analysis of MP16 - standard mass (MP0) + 16% kaolin residue.

It should be noted that for the dilatometry of MP32, Figure 37, there is a change in the curve, i.e. the slope of the curve rises again. This behaviour can be attributed to the considerable increase in kaolin added to the paste with the inclusion of 32 % residue, thus increasing the refractoriness of the paste.

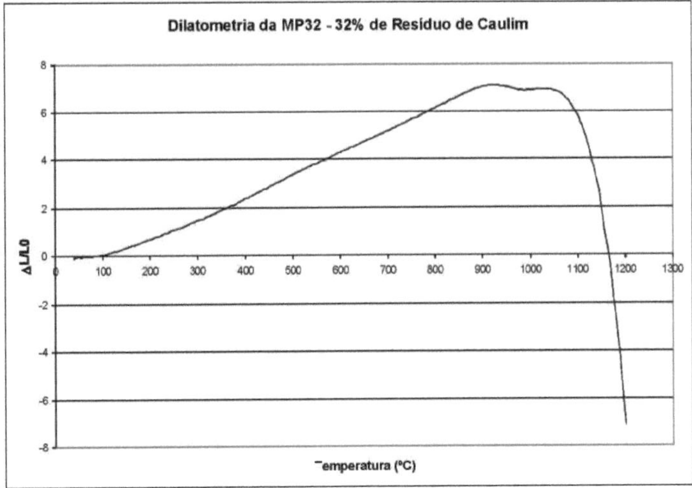

Figure 37 - Dilatometer analysis of MP32 - standard mass (MP0) + 32% kaolin residue.

8.4   TECHNOLOGICAL TESTS

The technological tests presented below refer to the samples sintered at the three study temperatures, with the values presented at 123C °C referring to the specimens sintered in industry. These values replaced the values obtained in the tests on the specimens sintered in the furnace at the LMC - UFRN, as the results were similar.

The linear shrinkage after firing is shown below, in Figure 38, for all the formulations containing kaolin residue (MP1 to MP32) and for the standard mass, MP0. The test specimens with kaolin residue in their mass show a similar trend over the three sintering temperatures, with a higher percentage of shrinkage at the lowest sintering temperature, from 8.58 to 9.13%, for the formulations with kaolin residue sintered at 1210 °C. This behaviour can be attributed to the greater loss of mass in these compositions, due to the decomposition of kaolinite, generating very reactive particles with high sinterability, as well as the large formation of liquid phase, also due to the presence of kaolin residue, which in addition to kaolinite, is rich in muscovite mica, which in turn has a high potassium (K_2O) content, which contributes effectively to the formation of liquid phase.

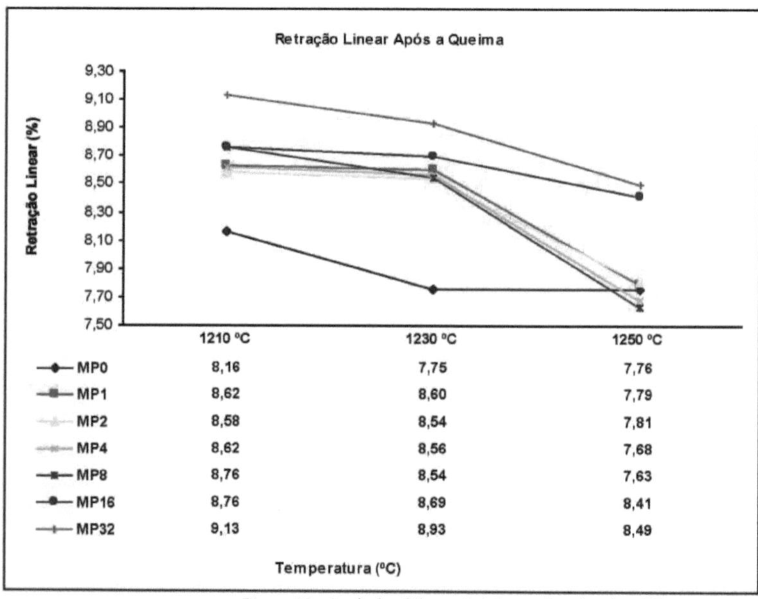

	1210 ºC	1230 ºC	1250 ºC
MP0	8,16	7,75	7,76
MP1	8,62	8,60	7,79
MP2	8,58	8,54	7,81
MP4	8,62	8,56	7,68
MP8	8,76	8,54	7,63
MP16	8,76	8,69	8,41
MP32	9,13	8,93	8,49

Figure 38 - Analysis of linear shrinkage after sintering (%).

This behaviour may also be an indicator that the sintering processes are taking place at a lower temperature than when using the standard mass, without the presence of kaolin residue, indicating the possibility of reducing the sintering temperature, which will lead to a reduction in energy costs in the industry and in the sintering cycle time, i.e. greater productivity.

On the other hand, the opposite is true at higher temperatures, i.e. the specimens shrink less. This reduction in the percentage of linear shrinkage may be directly linked to the reduction reactions that take place during the sintering process, causing the pieces to swell due to the trapping of gases released inside them during these reactions, increasing the volume of the sintered specimen and potentially damaging its mechanical strength and water absorption characteristics.

With regard to the MP32 formulation, despite showing the same behaviour as the other formulations containing kaolin processing residue, it does not meet one of the conditions required for the product in terms of linear shrinkage, as it shows a shrinkage of over 9%, which is the maximum allowed for single-colour products. This highlights the need to study formulations with kaolin residue contents of between 16% and 32%, as this can lead to exaggerated shrinkage, which is neither desired nor permitted by the standard.

Figure 39 shows the results of analysing the apparent porosity of the formulations. The results show that the compositions with kaolin residue sinter better at lower temperatures and thus have lower porosity values due to the high amount of glass phase promoted by the residue, probably due to the presence of muscovite mica which, together with the feldspar present in the standard mass, act as melters during the firing cycle, as observed in the dilatometric study of the masses. However, at higher temperatures the situation is reversed, i.e. there is an increase in the percentage of porosity, possibly due to the reduction reactions that can occur at higher temperatures. Porosity has a direct influence on the mechanical strength of the sintered material, so the lower the percentage of pores, the more resistant the body will be, which may be an indicator of the possibility of improving the standard mass with the addition of kaolin residue.

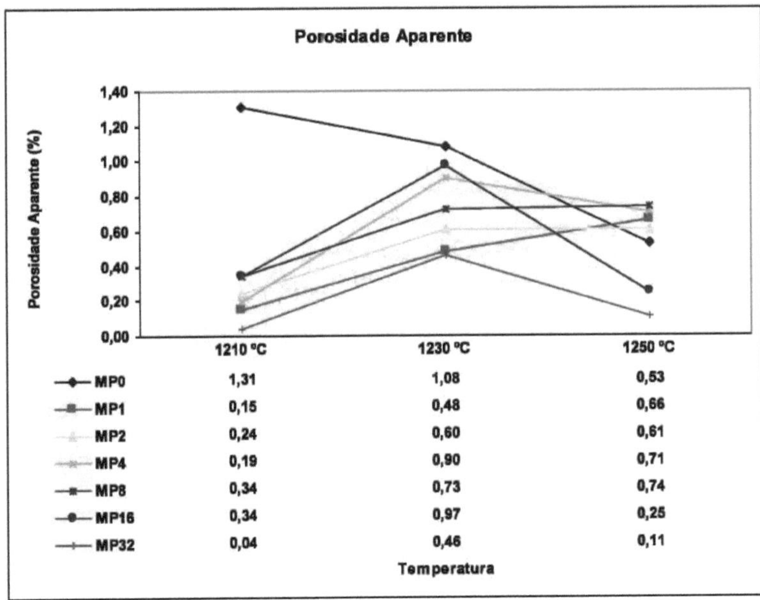

	1210 °C	1230 °C	1250 °C
MP0	1,31	1,08	0,53
MP1	0,15	0,48	0,66
MP2	0,24	0,60	0,61
MP4	0,19	0,90	0,71
MP8	0,34	0,73	0,74
MP16	0,34	0,97	0,25
MP32	0,04	0,46	0,11

Figure 39 - Apparent porosity analysis (%).

The result shown for water absorption, Figure 40, confirms the above, i.e. in situations where linear

65

shrinkage is lower and apparent porosity is higher, a higher percentage of water absorption is obtained, i.e. at higher temperatures, with the exception of MP0, which is the standard mass without the presence of residue, a high percentage of water absorption is observed, which is in line with the product found and tested on the market by the company in question. In the case of the lower temperature, the ceramic bodies with kaolin residue remained as expected, i.e. a lower percentage of water absorption, as explained above. The water absorption rates for the 1250 °C temperature are peculiar in that they are lower than those for the 1230 °C temperature. This is due to the great vitrification of the external surface of the specimens caused by the large amount of melting material present in the mixtures with kaolin residue.

This can be confirmed by looking at the XRD of the sintered sample, where there is an increase in the diffractogram line in relation to the abscissa axis, indicating a greater quantity of amorphous phases, part of which is the glassy phase generated by the dissolution of the alkali metals present in the raw materials. The micrographs of the sintered bodies, presented later, also confirm the above, where the microstructural evolution of the sintered body can be clearly seen.

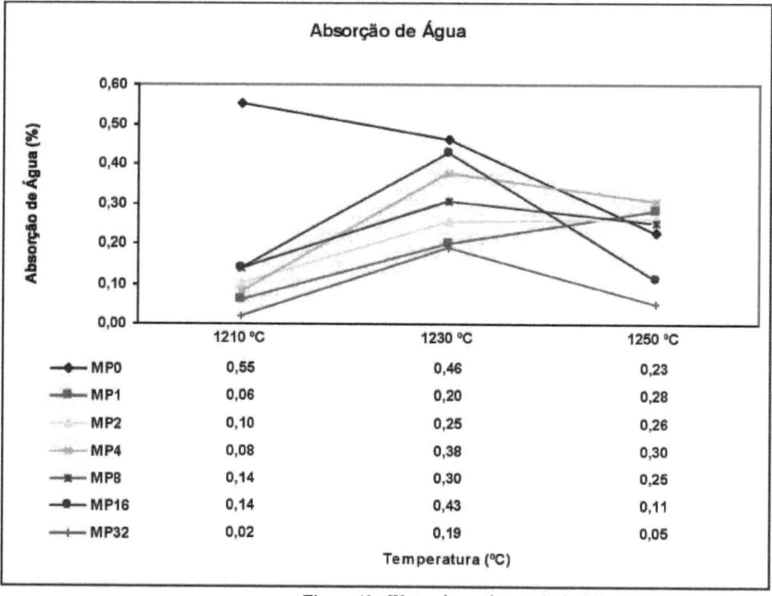

	1210 ºC	1230 ºC	1250 ºC
MP0	0,55	0,46	0,23
MP1	0,06	0,20	0,28
MP2	0,10	0,25	0,26
MP4	0,08	0,38	0,30
MP8	0,14	0,30	0,25
MP16	0,14	0,43	0,11
MP32	0,02	0,19	0,05

Figure 40 - Water absorption analysis (%).

The results of the apparent specific mass analysis (Figure 41) are intrinsically linked to the amount of pores in the sintered ceramic body. The results shown below show a gradual decrease in the apparent specific mass as the sintering temperature increases. This is true for all compositions containing kaolin residue, except for MP1 which, due to its small percentage, did not suffer a significant influence on its properties. Due to the increase in the amount of gas generated during the sintering of the test specimens at the higher temperatures and the trapping of these gases inside the ceramic body, there is a decrease in the specific mass, as the gases generated are unable to escape as a result of the closure of the open porosity and the high vitrification that occurs on the external face of the body, as mentioned above, increasing the volume of the sintered body and decreasing its mass. This behaviour can be confirmed by observing the increasing mass loss shown above in the thermogravimetric analysis curve of the kaolin residue, Figure 28, for temperatures above 1200°C. At 1210 °C, the specimens containing kaolin residue have a considerably higher apparent specific mass than the other temperatures. This is due to the presence of muscovite mica, present in the kaolin residue, which together with feldspar-K and albite present in the standard mass, MP0, promote the formation of a liquid phase. This flows into the interstices between the more refractory particles and, due to capillary forces, causes these particles to come closer together, resulting in quite significant shrinkage, as shown above. The greater the amount of material that melts and the lower the viscosity of the liquid formed, the greater the shrinkage and consequently the greater the density of the sintered material.

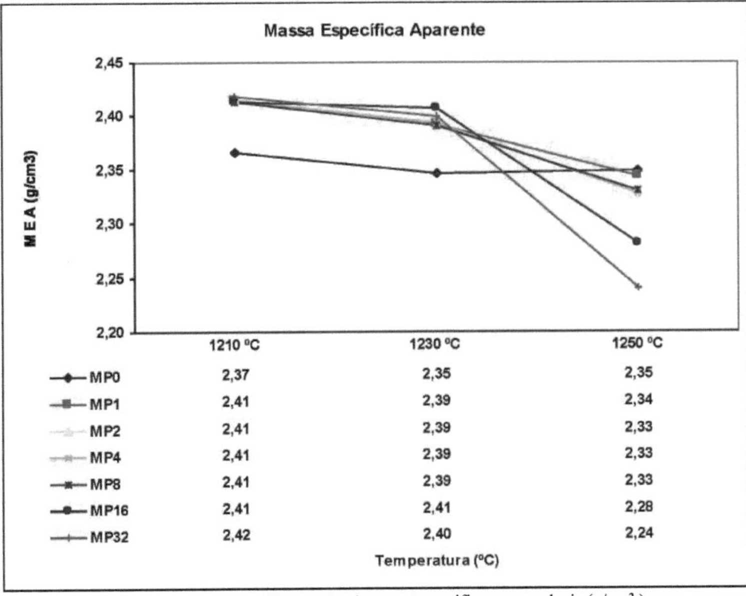

	1210 ºC	1230 ºC	1250 ºC
MP0	2,37	2,35	2,35
MP1	2,41	2,39	2,34
MP2	2,41	2,39	2,33
MP4	2,41	2,39	2,33
MP8	2,41	2,39	2,33
MP16	2,41	2,41	2,28
MP32	2,42	2,40	2,24

Figure 41 - Apparent specific mass analysis (g/cm²).

Figure 42 shows the behaviour of the three-point bending strength of the test specimens as a function of sintering temperature and the percentage of kaolin residue present in the mass. It can be seen that the formulations with kaolin residue in their mass show the best results when compared to MP0. Another point to note is the considerable difference in resistance between the highest and lowest sintering temperatures, which confirms the results shown so far. This behaviour may be associated with the large amount of glass phase present at the lower temperatures, and the formation of mullite promoted by the addition of kaolin residue. As the temperature rises, the secondary mullite crystals dissolve and the amount of amorphous phase and vitreous phase increases, making the ceramic body more fragile, as well as increasing the amount of pores inside the body, a fact proven earlier by analysing the apparent specific mass. It is believed that the reduction in strength is directly linked to the excess glassy phase produced at the three temperatures. This liquid phase is capable of dissolving part of the primary and secondary mullite crystals and also changes the nucleation and growth process of the secondary mullite crystals. As can be seen when analysing the micrographs of the test specimens later on.

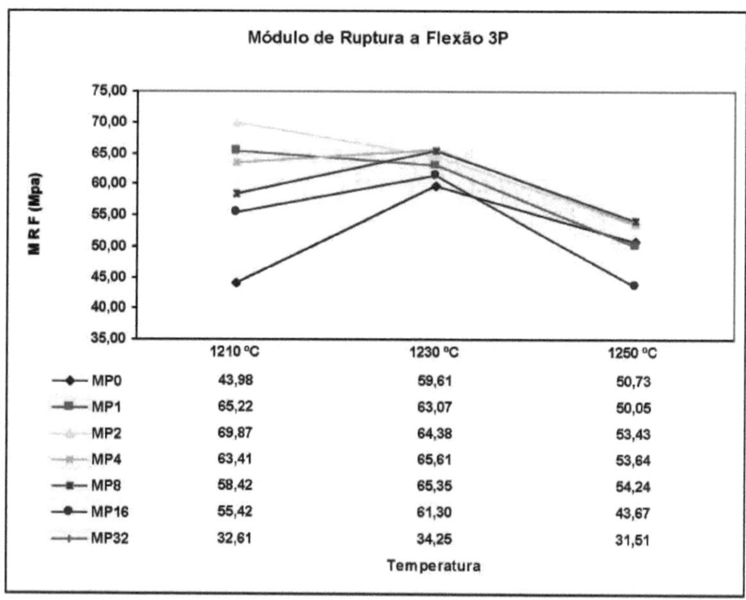

Figure 42 - Analysis of the 3-point flexural modulus of rupture in MPa.

As can be seen, the standard mass with the presence of kaolin residue showed a significant improvement in its properties compared to the standard mass without the residue when sintered at its temperature of 1230 °C, another fact that can also be observed is the improvement of these same properties in the bodies with the presence of kaolin residue when sintered at 1210 °C, this type of behaviour suggests a possible reduction in the sintering temperature in the industry.

8.5   STRUCTURAL AND MORPHOLOGICAL CHARACTERISATION OF SYNTERISED

8.5.1   DRX ANALYSIS

As the industry's production process uses a temperature of 1230 °C as the threshold for producing porcelain tiles, this temperature was analysed for comparison purposes. Therefore, the phase transformations that took place during the firing cycles were observed and the results compared, where it was found that the same phases found in the test specimens sintered in the ceramics industry's kiln are formed in the cycle adopted in the laboratory, which validates the process and the cycle adopted in the laboratory and makes the results reliable. Figures 43 to 47 show the results obtained from the mineralogical analysis by X-ray diffraction of the specimens sintered in the kilns at the UFRN ceramic materials laboratory - LMC and at the ceramics industry plant. The purpose of this analysis was to confirm whether the firing cycle adopted in the laboratory was forming the same phases found in the test specimens sintered in the industry's kiln.

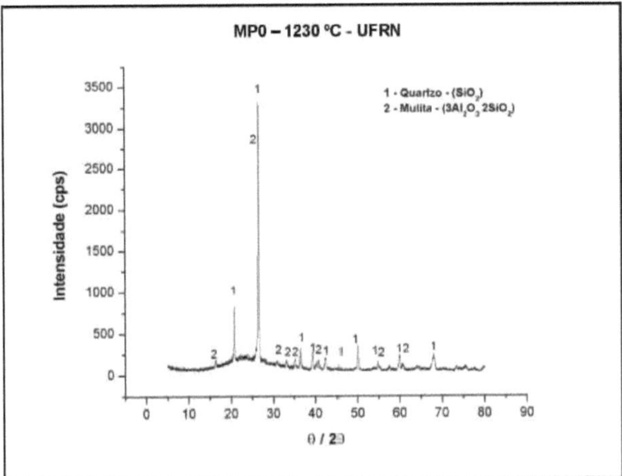

Figure 43 - Mineralogical analysis by XRD of MP0 sintered at 1230 °C in the furnace of the LMC - UFRN.

69

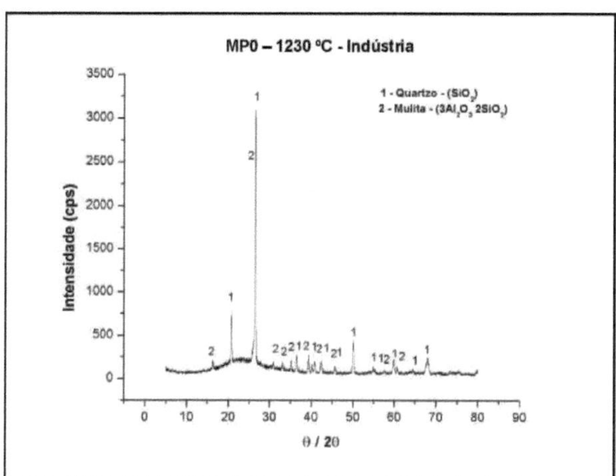

Figura 44   - Mineralogical analysis by XRD of MP0 sintered at 1230 °C in a ceramics industry kiln.

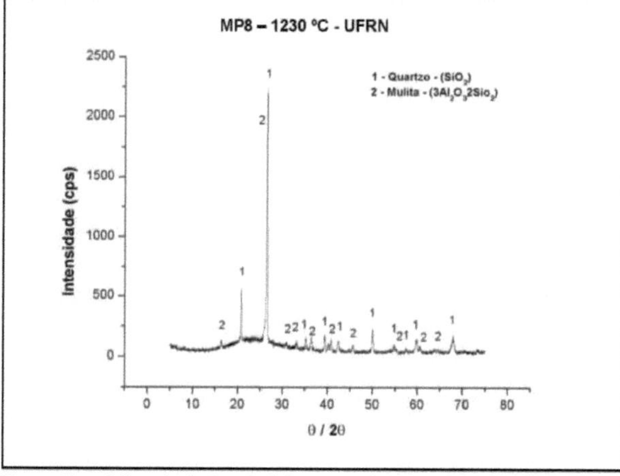

Figura 45   - Mineralogical analysis by XRD of MP8 sintered at 1230 °C in the furnace of the LMC - UFRN.

70

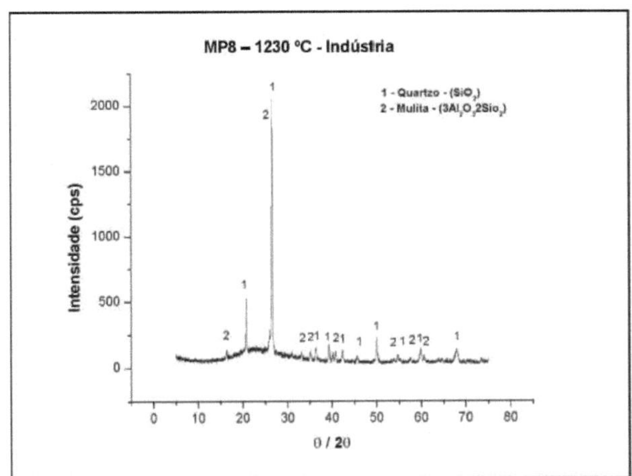

Figure 46 - XRD mineralogical analysis of MP8 sintered at 1230 °C in the ceramics industry kiln.

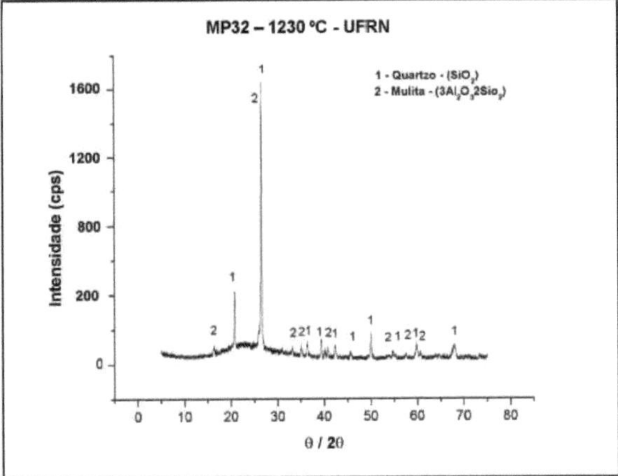

Figure 47 - Mineralogical analysis by XRD of MP32 sintered at 1230 °C in the furnace of the LMC - UFRN.

The phases identified by X-ray diffraction analysis are the same as those found for the two sintering situations, i.e. both in the ceramics industry kiln and in the LMC - UFRN kiln, quartz and mullite were identified, the latter coming from the kaolin in the standard mass and the kaolin present in the residue added to it. This fact proves that even in the composition with the greatest presence of kaolin residue, MP32, its existence did not hinder or prevent the formation of the main phases in the final product, but rather contributed to greater formation of the vitreous phase and mullite, consequently increasing the apparent specific mass and mechanical strength of the final product when compared to the standard mass (MP0), in compositions with up to 16% kaolin residue added. Mullite is formed from spinel and amorphous aluminosilicate, obtained at 985 °C from clays such as kaolinite, illite and other aluminosilicates as a result of the first liquid phase formation, which also occurs at around this same temperature, the eutectic temperature of the feldspar-K + SiO2 mixture, and which provides the reaction of the liquid formed reacts with SiO2 eliminated from the metakaolinite. The Al2O3 and SiO2 from the metakaolinite are transformed into primary mullite with very small lamellar crystals, which appear in aggregates where there was previously kaolinite, which can be confirmed by microstructural analysis. As the temperature continues to rise, from around 1150 °C to 1250 °C, the primary mullite aggregates are penetrated by alkaline ions from feldspar and muscovite mica and the depletion of alkalis from the K2O - Al2O3 - SiO2 system allows secondary mullite to crystallise.

It can also be seen that there is an increase in the diffractogram line in relation to the abscissa axis, while some of the peaks in the diffractograms of the raw materials no longer exist. This indicates a greater quantity of amorphous phases, part of which is the glassy phase generated by the dissolution of the alkali metals present in the raw materials. Another point to consider is that the sintering reactions are taking place at lower temperatures than expected (1230 °C), which makes it possible to reduce the sintering temperature and the firing level during the industrial process.

This characteristic is made clear in Figures 48 to 50 below, which show some of the X-ray diffractograms of the ceramic masses sintered at 1210 °C in the LMC - UFRN furnace. These diffractograms show that the phases present in the final product are the same as those found in the diffractograms of those sintered at 1230 °C, i.e. quartz and mullite, although the intensities of the characteristic peaks are lower than these. This situation may directly reflect the difference between the degree of crystallisation of the mullite formed at the temperatures mentioned above. This could have a considerable influence on the physical and mechanical properties of the sintered material.

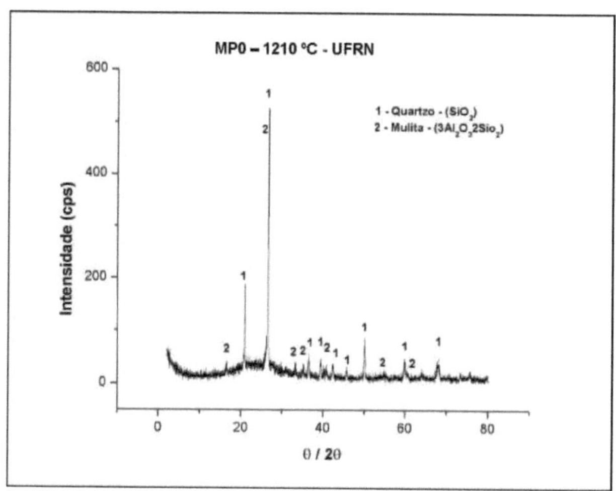

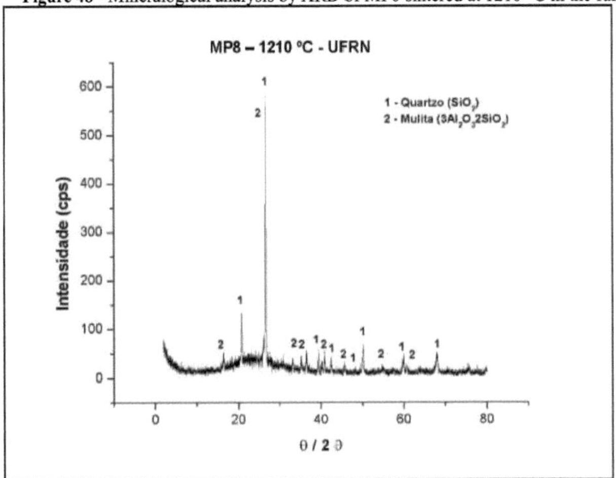

Figure 48 - Mineralogical analysis by XRD of MP0 sintered at 1210 °C in the furnace of the LMC - UFRN.

Figure 49 - Mineralogical analysis by XRD cf MP8 sintered at 1210 °C in the furnace of the LMC - UFRN.

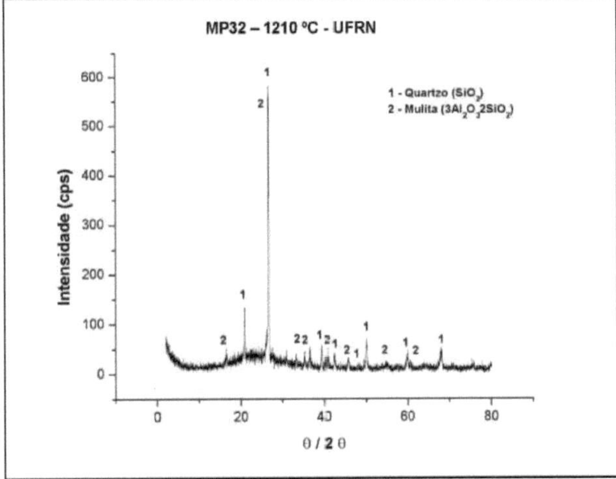

Figure 50 - Mineralogical analysis by XRD of MP32 sintered at 1210 °C in the furnace of the LMC - UFRN.

The X-ray diffractogram of the MP0 formulation, a standard mass without the presence of kaolin processing residue, at 1210 °C also shows the same phases present in the diffractograms of the masses with the presence of kaolin processing residue. However, in order to confirm that a possible lowering of the sintering temperature can be carried out, even without the presence of the residue in the paste, a microstructural analysis of the sintered materials was carried out using scanning electron microscopy.

8.5.2 MICROSTRUCTURAL CHARACTERISATION BY SEM

The microstructural characterisation of the sintered specimens was carried out in two stages:

- Fracture surface analysis and

- Analysis of the polished and attacked surface.

8.5.2.1   MICROSTRUCTURAL ANALYSIS OF THE FRACTURE SURFACE

The fracture surface was analysed in order to verify the microstructure formed and any significant difference between the different formulations and firing cycles adopted. The micrograph below, Figure 51, refers to the standard mass without the presence of kaolin residue, MP0.

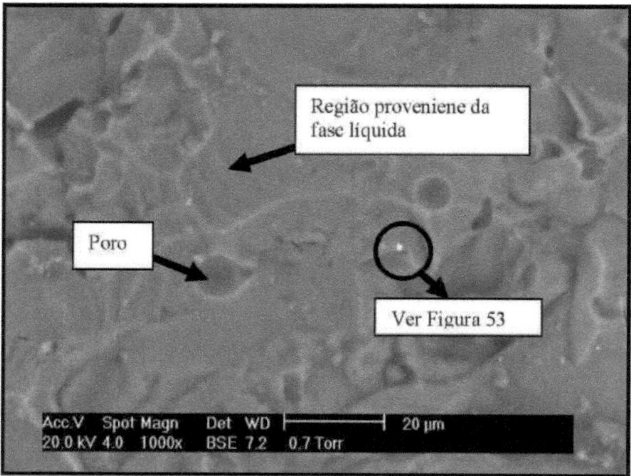

Figure 51 - SEM micrograph of MP0 sintered at 1230 °C in the ceramics industry kiln, 1000X.

When comparing the micrograph shown in Figure 51, referring to the standard mass without the addition of kaolin residue (MP0), sintered at 1230 °C, with the one shown in the micrograph in Figure 52 (MP8), it can be seen that MP0 has larger and more irregularly shaped pores, which indicates that simply lowering the sintering temperature without the presence of kaolin processing residue could jeopardise the technical characteristics of the final product, to be observed in the technological tests carried out and presented above.

Figure 52 - SEM micrograph of MP8 sintered at 1230 °C in the ceramics industry kiln, 1000X.

Figure 53, EDS, shows the presence of remnants of feldspar left over from the standard mass, since the residue does not have this phase present in its X-ray diffractogram. From a chemical point of view, it can be a source of staining, fluorescence and expansion due to humidity, as it has a high concentration of potassium (K) and sodium (Na), which are hygroscopic. This was confirmed by the company itself.

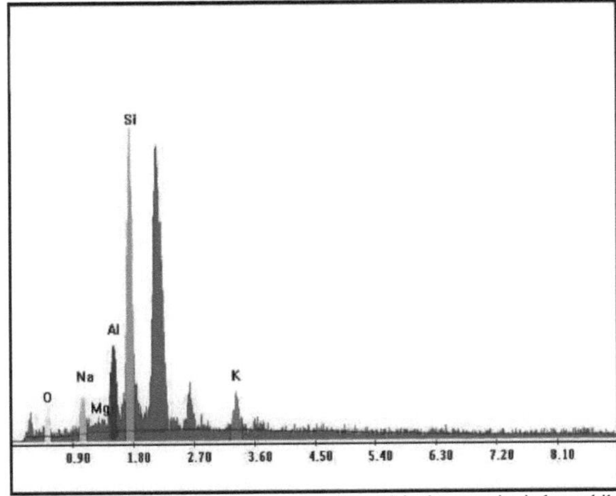

Figure 53 - SEM EDS of MP0 sintered at 1230 °C in the ceramics industry kiln.

As can be seen, with the addition of kaolin residue to the standard mass, the amount of glassy phase increases and consequently the pores become smaller. This behaviour is due to the increase in the melting material, muscovite mica, which is present in considerable quantities in the kaolin residue, as shown above in Table 5. An important observation is that this type of behaviour is repeated for the temperature of 1210 °C, Figure 54, which may suggest a possible reduction in the sintering temperature without affecting the properties of the final product.

Figure 54 - SEM micrograph of MP8 sintered at 1210 °C in the furnace of the LMC - UFRN, 1000X.

8.5.2.2 MICROSTRUCTURAL ANALYSIS OF THE POLISHED AND ATTACKED SURFACE

The microstructural evolution of the test specimens (SPs) sintered at various temperatures and with the addition of kaolin residue ranging from 0 to 32% by weight, MP0 to MP32 respectively, was analysed. After polishing, the CPs were etched with 2% hydrofluoric acid (HF) in water. HF etching was used to partially remove the glass phase and consequently reveal the crystalline phases present in the samples more easily. In general, evidence of mullite formation was found in the various processing conditions.

In general, the samples formed by the standard composition revealed a microstructure made up of quartz grains and primary mullite surrounded by a glassy phase. The results obtained by X-ray diffraction (Figure 48) confirm the presence of these phases. A detailed microstructural examination of the standard samples was carried out, and in a few places in the microstructure it was possible to identify the formation of secondary mullite. The micrograph in Figure 55, referring to the standard sample, MP0, treated at 1210° C, illustrates one of the rare microstructural occurrences in which it was possible to identify the presence of secondary mullite. In this case, secondary mullite formations are always associated with a massive formation of primary mullite crystals. This type of microstructural occurrence is typical of systems that develop little liquid phase or are subjected to a very short firing stage. In these systems, the presence of secondary mullite is observed strictly in regions of internal pores with an interface between the primary mullite and the liquid phase formed by the fusion of feldspars. Figure 55 illustrates this type of occurrence.

Figure 55 - SEM micrograph of MP0 sintered at 1210 °C in the furnace of the LMC - UFRN, 5000X.

The micrograph in Figure 56 is a magnification of the highlighted region in the previous micrograph, Figure 55. This micrograph clearly shows the formation of acicular crystals of secondary mullite. The chemical analysis by EDS (Figure 57) was carried out at specific points in this region of the sample. The three points analysed showed the presence of silicon as the majority element and aluminium as the minority. It can also be seen that the amount of aluminium increases slightly from position A to C and from C to B. This result is an indication that the predominant primary mullite in region C is being dissolved by the surrounding glassy phase and recrystallising in the form of acicular crystals of secondary mullite. Point A corresponds to a region abundant in silica-rich glassy phase.

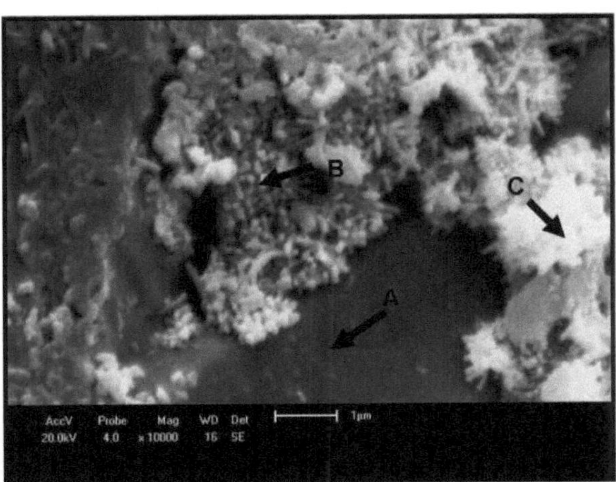

Figure 56 - SEM micrograph of MP0 sintered at 1210 °C in the furnace of the LMC - UFRN, 10000X.

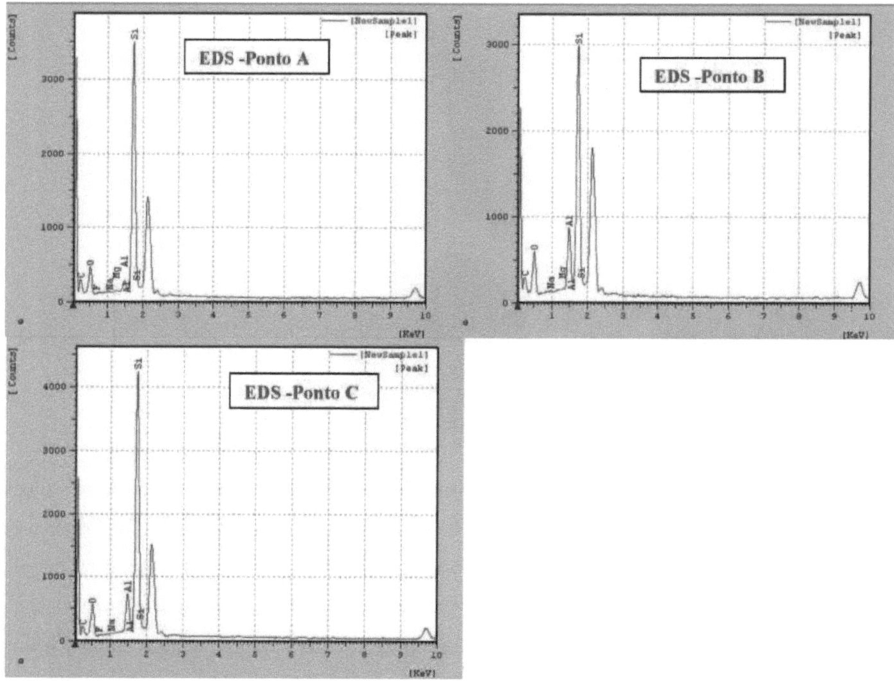

Figure 57 - EDS of MP0 sintered at 1210 °C in the furnace of the LMC - UFRN.

Figure 58 corresponds to the micrograph of the standard formulation, sintered at a temperature of 1230 °C. It can be seen that with an increase of 20° C, there is a slight increase in the amount of secondary mullite.

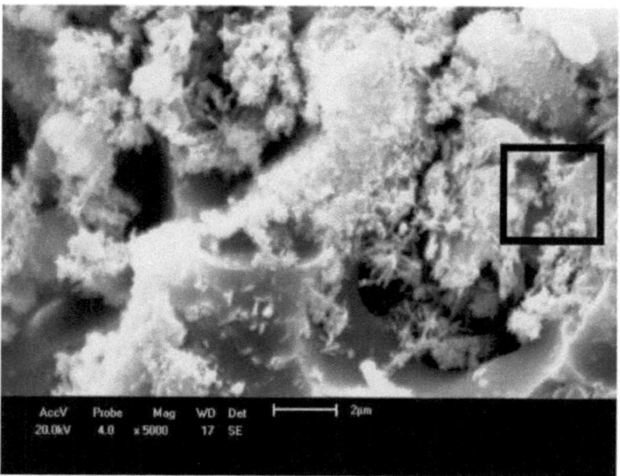

Figure 58 - SEM micrograph of MP0 sintered at 1230 °C in the ceramics industry kiln, 5000X.

Figure 59 is an enlargement of the region shown in Figure 58. The results of the chemical microanalysis show that the intensities of the aluminium and silicon peaks are in agreement with the

78

corresponding type of morphology at each point analysed. Thus, point A, which qualitatively showed the least amount of aluminium, corresponds to a region abundant in a glassy phase rich in silica. Point B shows an increase in the amount of aluminium and the presence of acicular crystals associated with the formation of secondary mullite. At point C there is an intermediate situation between the two cases, with massive formation of primary mullite and a vitreous phase rich in aluminium, from which secondary mullite crystals nucleate and grow (Figure 60).

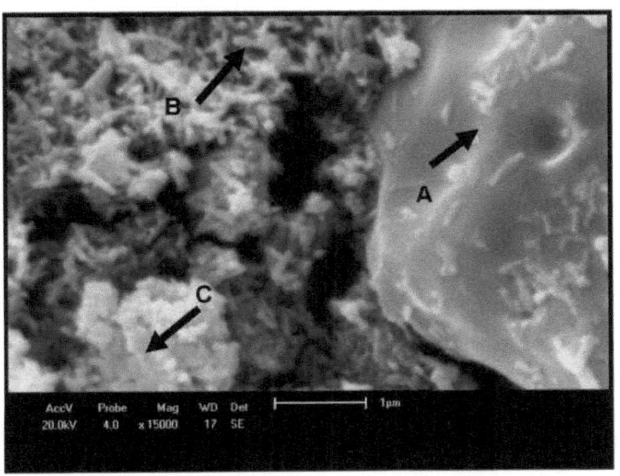

Figure 59 - SEM micrograph of MP0 sintered at 1230 °C in the ceramics industry kiln, 15000X.

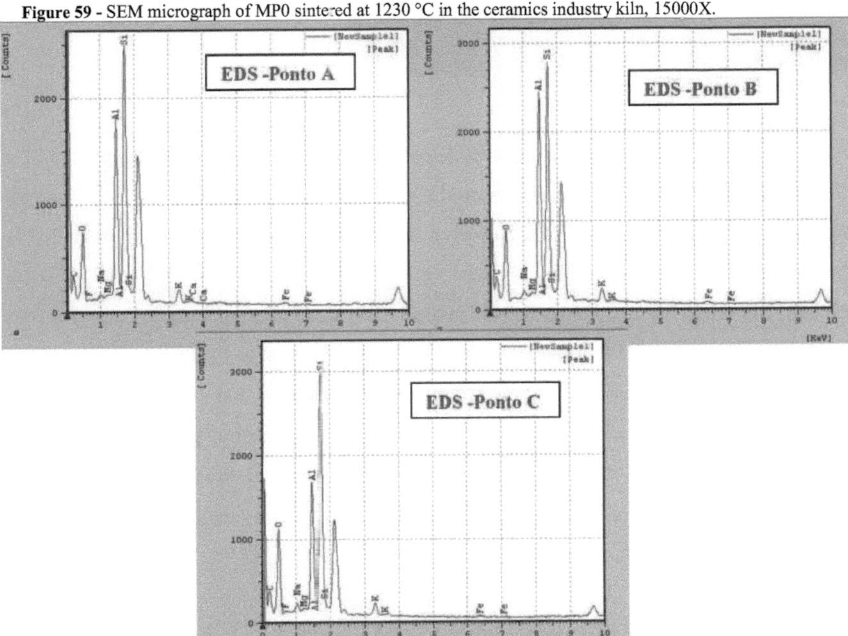

Figure 60 - EDS of MP0 sintered at 1230 °C in the ceramics industry kiln.

Microstructural characterisation of the entire series of samples containing residue made it possible to assess the effect of residue on microstructural evolution.

To better emphasise the microstructural evolution observed, the results corresponding to the 2%, 4% and 16% residue contents will be presented.

Figure 61 shows the micrograph of sample MP2, sintered at 1210° C. The results show that the sample treated at 1210° C had a microstructure with better formation of secondary mullite crystals compared to the sample without residue. Therefore, the presence of kaolin residue contributes to improving the formation of secondary mullite. This result can be attributed to the increase in the amount of glassy phase provided by the introduction of mica into the system.

Figure 61 - SEM micrograph of MP2 sintered at 1210 °C in the furnace of the LMC - UFRN, 5000X.

On the other hand, increasing the sintering temperature to 1230° C led to the formation of a microstructure with different characteristics, as can be seen in Figure 62 (MP2 - 1230° C). You can see the presence of acicular mullite crystals. However, they are larger and have a morphology that is not as characteristic of secondary mullite crystals, i.e. imbricated needle-like. This result may be associated with the presence of mica associated with the increase in temperature, leading to the formation of a greater quantity of liquid phase. This liquid phase is capable of dissolving part of the primary and secondary mullite crystals and also changing the nucleation and growth process of the secondary mullite crystals. As can be seen above in the results of the technological tests relating to the three-point flexural modulus of rupture, the decrease in strength shown in Figure 42 for the MP2 formulation when the temperature is raised from 1210 °C to 1230 °C may be directly linked to this type of microstructural formation.

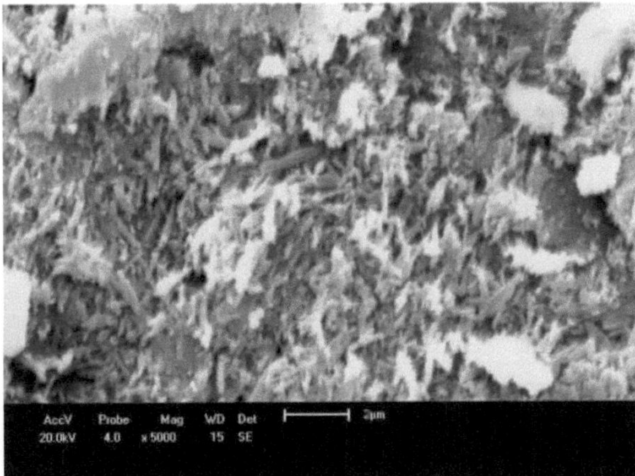

Figure 62 - SEM micrograph of MP2 sintered at 1230 °C in the ceramics industry kiln, 5000X.

Figure 63 contains the micrograph corresponding to the MP4 sample sintered at a temperature of 1210° C. The results show that increasing the amount of residue contributes to increasing the formation of secondary mullite. This can be clearly seen in this micrograph.

Figure 63 - SEM micrograph of MP4 sintered at 1210 °C in the furnace of the LMC - UFRN, 5000X.

Figure 64 shows an enlargement of the region shown in Figure 63. The results confirm the presence of intense formation of acicular crystals of secondary mullite. The points A and B indicated correspond to the EDS spectra shown in Figure 65. The EDS results are in line with the morphology of the regions analysed. Thus, in point A it is clear that the greater amount of aluminium is associated with the presence of secondary mullite, while point B shows a greater amount of silica and a morphology characteristic of a glassy phase.

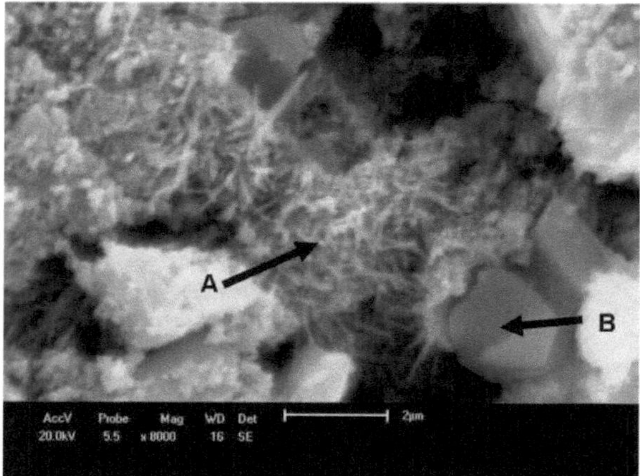

Figure 64 - SEM micrograph of MP4 sintered at 1210 °C in the furnace of the LMC - UFRN, 8000X.

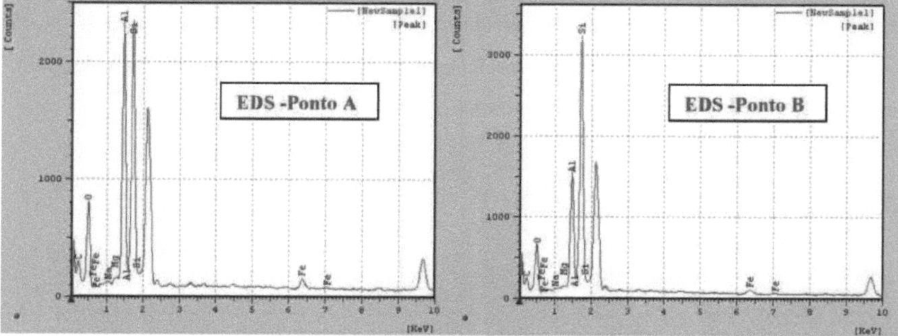

Figure 65 - EDS of MP4 sintered at 1210 °C in the furnace of the LMC - UFRN.

Figure 66 shows the micrograph corresponding to the MP4 sample sintered at a temperature of 1230° C. The results show that the increase in the amount of residue associated with the increase in sintering temperature causes a reduction in the formation of secondary mullite, this behaviour can be associated with the increase in the amount of liquid phase, due to the higher residue content and the increase in sintering temperature.

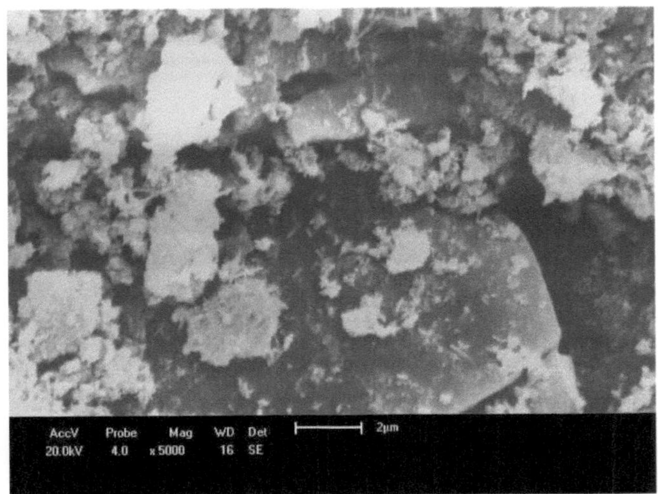

Figure 66 - SEM micrograph of MP4 sintered at 1230 °C in the ceramics industry kiln, 5000X.

Figure 67 shows the micrograph corresponding to sample MP16 sintered at a temperature of 1210° C. In this case, the presence of secondary mullite was not verified, but only the indication of primary mullite surrounded by a glassy phase.

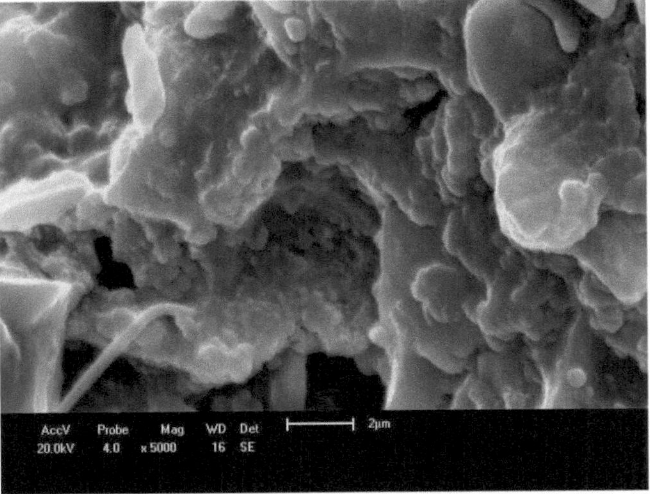

Figure 67 - SEM micrograph of MP16 sintered at 1210 °C in the furnace of the LMC - UFRN, 5000X.

Figure 68, MP16 sintered at 1230 °C, shows that the morphology of the microstructure undergoes a marked change, i.e. the formation of massive crystals of primary mullite.

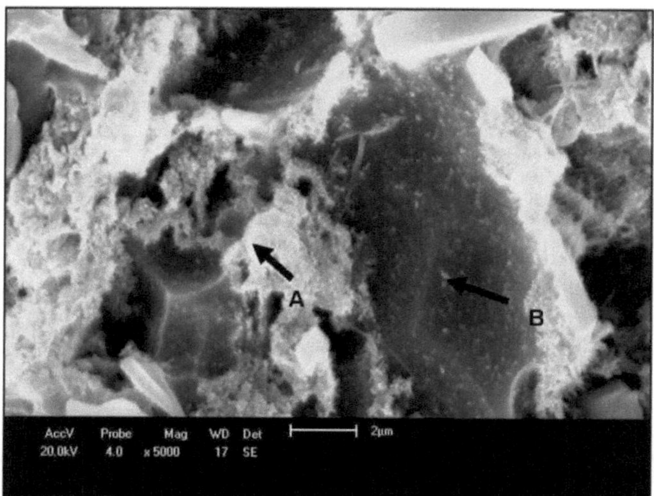

Figure 68 - SEM micrograph of MP16 sintered at 1230 °C in the ceramics industry kiln, 5000X.

Figure 69 contains the results of the chemical microanalysis by EDS at points A and B indicated in Figure 68. The results confirm the statements made based on analysing the morphology of sample MP16.

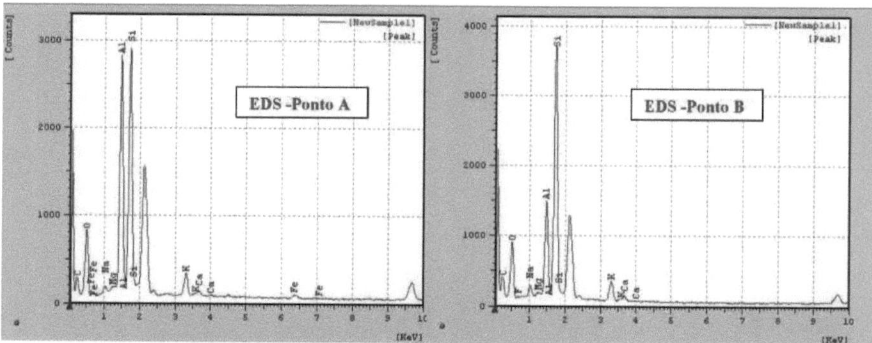

Figure 69 - EDS of MP16 sintered at 1230 °C in the ceramics industry kiln.

The joint analysis of the microstructural evolution results can help to understand the mechanical behaviour results presented in section 8.4 technological tests.

The results shown in Figure 42 indicate that at temperatures of 1210° C and 1230° C the addition of waste always improves the flexural strength of the compositions studied. This behaviour could be attributed to the formation of more liquid phase due to the presence of mica, which causes a reduction in porosity, as shown in Figure 39. However, the results of the microstructural analysis indicate that the presence of residue contributes to increasing the presence of mullite with an acicular crystalline habit. This type of crystal occurs preferentially dispersed in the vitreous phase and can act as a reinforcement of it, behaving like a ceramic composite. In this way, the presence of secondary mullite improves the mechanical properties of the glass matrix.

CHAPTER 9

CONCLUSIONS

Based on the results shown above, it can be seen that the values suggested by Standard NBR 13818 - Ceramic tiles - Specification and Test Methods, were exceeded in the various formulations proposed in this thesis, where it can be concluded that the introduction of kaolin residue to the standard mix promotes significant improvements in the final product, with no need for any kind of change in the process already adopted by the company. It's just a simple addition of the residue, with its particle size and percentage controlled. In this way, the technical feasibility of using kaolin processing residue as a complementary raw material with great potential in the porcelain tile industry is understood.

According to the results obtained, it was possible to use kaolin residue in a percentage of up to 16% added to the standard mix, MP16. This formulation showed results compatible with those required by the current standard, NBR 13818/1997 - Ceramic tiles for cladding - specification and test methods, as well as reducing the cost of the mass and the cost of energy during the firing cycle. The three-point flexural modulus of rupture found at 1210 °C (55.4 MPa) was lower than that obtained when sintering the same MP16 formulation at 1230 °C (61.3 MPa). However, the standard in force requires 35 MPa, which is much lower than the lowest value. For the other tests, the values obtained for this formulation sintered at 1210 °C are always better than the results at 1230 °C.

As a result of its use in industry, the liabilities of the company that generates it are reduced and the costs of obtaining raw materials for the ceramics industry are minimised, thus helping to reduce the social and environmental damage caused by the exploitation and processing of ceramic raw materials.

It is worth noting that, due to the presence of highly melting phases in the raw materials, these values were accompanied by an average shrinkage after sintering of 8.6%, up to the percentage of 8% added kaolin residue, which is within the expected values for single colour products, which is between 8.5 and 9.0%. The technological properties required by the standard were achieved - water absorption, apparent specific mass and mechanical resistance - in all the aspects analysed. In other words, the presence of the residue kept these characteristics within acceptable values.

Microstructural analyses of the sintered bodies confirmed that all the phases required in the final product were present in the masses produced with the kaolin residue, which did not cause any damage to their formation.

When all the results of these tests and analyses are put together, the trend towards a lower sintering temperature is confirmed, due to the maturing of the mass before the temperature currently used in the sintering process in the industry, which means that in the formulation containing the kaolin residue, This indicates a possible decrease in sintering temperature, which will lead to a reduction in energy costs in the industry and in the sintering cycle time, and as a consequence of the latter, an increase in the company's productivity.

In addition to the above, these results also provide another guideline to be followed: reducing the percentage of feldspar in the standard mass, since mica is directly contributing to the formation of the liquid phase, acting as a flux, which is the function of the feldspar present in this standard mass. Reducing the percentage of feldspar in the standard mass would make the mass cheaper because it is an expensive raw material, unlike the kaolin processing residue. An important fact is that the kaolin residue, which is still discarded by the generating companies as a useless material, can be seen by them as a promising raw material for the ceramic floor and wall tile industry, either as a fluxing material, due to the large amount of muscovite mica present in its constitution, or as a structure-forming material due to the kaolin, in addition to promoting a clear firing, which increases the added value of the final product. This will make a positive contribution to reducing the negative impacts caused by the disorderly disposal of this material, as well as helping to reduce companies' liabilities.

REFERENCES

Abreu, S.F. - Mineral Resources of Brazil. Edgard Ltda, vol.I,1973.

Associação Brasileira de Normas Técnicas, NBR 13818/1997 - Ceramic tiles for cladding - specification and test methods.

Aléssio, S.R.T. - Effect of the Date of WTP Sludge Collection on the Physical Properties of Ceramic Specimens with Incorporated Sludge, 49th Brazilian Congress of Ceramics, 2005.

Albero, J.L.A. - The pressing operation: technical considerations and its industrial application. Part I: Filling the mould cavities. Revista Cerâmica Industrial, v.5(5), p. 23-28, 2000.

. - The pressing operation: technical considerations and its industrial application. Part II: Compacting. Industrial Ceramics Magazine, v.5(6), p. 14-20, 2000.

Alboneti, V. - Printing by continuous machining: l'impianto MTC41. - Ceramica Informazione 348, p. 177, 1995.

National Association of Ceramic Tile Manufacturers (Anfacer) - <http://www.anfacer.com.br> - accessed on 15 July 2006.

Brazilian Ceramics Association (ABCeram) - <http://www.abceram.org.br> - accessed on 25 July 2005.

AmbienteBrasil-
<http://www.ambientebrasil.com.br/composer.php3?base=residuos> - accessed on 03/06/05.

Arantes, F.J.S.; Galesi, D.F.; Quinteiro, E.; Boschi, A.O. The staining and closed porosity of porcelain stoneware. Cerâmica Industrial, v.6, 2001.

Barata, M.S. High Performance Concrete in Pará: study of the technical and economic feasibility of producing high performance concrete with materials available in Belém through the use of active silica and metakaolin additions.

1998. 164 f. Dissertation (Master's in Civil Engineering) - Postgraduate Programme in Civil Engineering, Federal University of Rio Grande do Sul, Porto Alegre.

Barata, M.S., Dal Molin, D.C.C. Preliminary evaluation of kaolinite residue from kaolin processing industries as a raw material in the production of a highly reactive metakaolinite - Ambiente Construído, Porto Alegre, v. 2, n. 1, p. 69-78, jan./mar. 2002.

Beltrán, V.; Ferrer, C.; Bagan, V.; Sanchez, E.; Garcia, J.; Mestre, S. Influence of pressed powder characteristics and firing temperature on the porous microstructure and stain resistance of porcelain stoneware tiles. Cerâmica Informação, v.2/3, 2000.

Berther, G - Cutting porcelain stoneware, technology, production, market. Gruppo Editoriale Faenza Editrice, 1993.

Biffi, G. - Porcelain stoneware - manufacturing manual and application techniques, Faenza Editrice do Brasil Ltda. São Paulo, p. 32, 2002.

Biffi, G.; Mazzacani, P. - Manuale per la produzione delle piastrelle. Gruppo Editoriale Faenza Editrice, 1997.

Borlini, C.; Mendonça, J.L.C.C.;. Vieira, C.M.F.; Monteiro, S.N - Effect of Sugarcane Bagasse Ash Granulometry on the Properties of a Kaolinitic Clay. 49th Brazilian Ceramics Congress, 2005.

Boschi, A. - Firing of Ceramic Bodies,<www.centraldaceramica.com.br> accessed on 14/11/2005.

Casagrande, M. C.; Novaes, A. P.; Hotza, D.; Alexandre, L. R.; Magagnin, F. S.; Sartor, M. N.; Montedo, O. R. K.- Recycling of Chamotte in the Ceramic Mass of Gresified Pavements, 2000.

Contoli, L.; Bresciani, A. - Porcelain stoneware taxes, technical and economic considerations - 5th Symposium of the S. Schmidt Group, Limburg, 1996.

Deer, W. A.; Howie, R. A.; Zussman - An Introduction to the Rock-Forming Mineral, Logman Group Ltd, p. 340- 355, London, 1975.

Figueiredo Gomes, C. S- What clays are and what they are used for. Calouste Gulbekian Foundation, Lisbon, 1986.

Fontenelle, A. P. G.; Abidack, R. C. - Cerâmica para revestimentos, BNDES Setorial, Rio de Janeiro, n. 10, p. 201-252, Sep. 1999.

Gilbertoni, C.; P. I. Paulin F.; M. R. Morelli - Characterisation of ceramics sintered by viscous flow, Revista Cerâmica, 51, p. 331-335, 2005.

Godinho, K.O.; De Holanda, J.N.F.; Da Silva, A.G.P- Effect of the Addition of Glass on the Firing Properties of a Red Clay. 49th Brazilian Ceramics Congress, 2005.

Harben, P. W. - The industrial minerals handbook. London: Industrial Minerals, Divison,. 253p., 1995.

Heck, C. - Porcelain stoneware, Revista Cerâmica Industrial, vol. 01, p. 04-05, August/December, 1996.

Albaro, J. L. Amorós - The Pressing Operation: Technical Considerations and its Industrial Application Part II: Compaction, Revista Cerâmica Industrial, vol. 05, p. 14-20, November/December, 2000.

Kobayashi, Y.; Ohira, O.; Satoh, T.; Kato, E. - Effect of quartz on the sintering and bending strength of the porcelain bodies in quartz-feldspar-kaolin system. J. Ceram .Soc. Jap. Int.Ed., 102(1), 100-105, 1994.

Llorens, F.G. Melting raw materials for the manufacture of porcelain stoneware. Cerâmica Informação, v.9, 2000.

Mello, M.; Gauita, J. P.; Fajan, S. F - Evaluation of the Possibility of Using Mud from the Marble Processing Process as a Raw Material in Red Ceramics. 49th Brazilian Ceramics Congress, 2005.

Menegazzo, A.P.M.; Lemos, F.L.N.; Paschoal, J.O.A.; Gouvêia, D.; Carvalho, J.C.; Nóbrega, R.S.N. Porcelain stoneware. Part I: a marketing approach. Cerâmica Industrial, v.5, 1-10, September/October, 2000.

Menegazzo, A.P.M.; Paschoal, J.O.A.; Andrade, A.M.; Carvalho, J.C.; Gouvêa, D. - Evaluation of the Mechanical Resistance and Weibull Modulus of Porcelain and Granite Stoneware Products. Cerâmica Industrial, v.7, 2002.

Mothé Filho, H.F.; Lemos, O.; Polivanov, H. - Recycling of Sand Extraction Tailings from the Itaguaí/Seropédica Region, RJ. 49th Brazilian Ceramics Congress, 2005.

Novaes, A.P. - Porcelain stoneware: market and technological aspects. Revista Cerâmica Industrial, v.3, 34-41, May/June, 1998.

Nolasco, A. P. - Votorantin Waste Becomes Brick Input, <www.unilivre.org.br/banco de dados> accessed on 29 July 2005.

Orts, T. M. J. - Sinterisation of sandblasted floor tiles. Castellón: Universitat de València, 1991. Doctoral thesis.

Palmonari, C. - II porcellanato stoneware - Edizione Castellarano Fiandre Ceramiche SpA, Ceramic Centre, Bologna 1989.

Pinheiro, P.G.; Fabris, J.D.; Mussel, W.N.; Murad, E.; Scorzelli, R.B. - Chemical characterisation, occurrence and distribution of iron in kaolin minerals from the region of Mar de Espanha (MG) <http://www.ufmg.br/prpg/dow anais>.

Rado, P. - An introduction to the technology of pottery. 2nd. ed. Oxford: Pergamon Press, 1988. Chapter 6. Firing, p. 92-97.

Ramalho, M.A.F.; Almeida, R.R.; Neves, G.A.; Santana, L.N.L. - Study of the Potential of Kaolin and Granite Waste for the Production of Ceramic Blocks. 49th Brazilian Ceramics Congress, 2005.

Reed, J. S. - Principles of ceramics processing. 2. ed. New York: John Wiley & Sons, 1995.

. - Desde la carga hasta la baldosa prensada: Mecánica y cambios microestructurales del sistema. Proceedings of Qualicer 2000, vol. I, p. Con-23-41 (2000).

Federative Republic of Brazil - Ministry of Mines and Energy Secretariat of Mines and Metallurgy Brasília. Pegmatites of the Northeast: Diagnosis on rational and integrated utilisation - 2003.

CONAMA Resolution no. 307, of 5 July 2002.

HereVip Magazine
<http://www.revistaaquivip.com.br/noticias displays.php?rel=MTIx&id=MTIy > accessed on 05/01/2007.

Ribeiro, M.J.P.M.; Labrincha, J.A.; Ventura, J.M.G. - Atomisation: Influence of some of the process variables. Kéramica, n. 237, p. 18-28 (1999)

Rocha, C. V. ; Cardoso, A. V. - Preparation and characterisation of glass based on blast furnace slag. In: BRAZILIAN CONGRESS OF CERAMICS, 43, 1999, Florianópolis. Proceedings... São Paulo: ABCERAM, (1999), 257, 01-12.

Sacmi - Fine porcelain stoneware - Edizione Sacmi Imola 1996.

Sane, S.C.; Cook, R.L. - Effect of grinding and firing treatment on the crystalline and glass content and the physical properties of whiteware bodies. J. Am. Ceram. Soc. 34(5), 145-151, 1951.

Sánchez, E.; Orts, M.J. ; García-Ten, J.; De Lemus, R. - Effect of porcelain tile raw materials composition on the arising phases in firing. 9th Symposium International on Ceramics. SIMCER. Bologna (Italy), 5-8 oct. 1998.

Sainz, J.G.; Ripollés, R.R. - Laboratory controls for porcelain stoneware. Ceramic Information, v.5, 1999.

Souza Santos, P. - Ciência e Tecnologia de Argilas, São Paulo: Editora Edgard Blucher Ltda, 1989.

Teixeira, R.; Souza, A. E.; Santos, G. T. A.; Dias, F. C. - Characterisation of Clays Used by the Red Ceramic Industry of Martinópolis - SP and Effect of the Incorporation of Residues on their Ceramic Properties. 49th Brazilian Ceramics Congress, 2005.

Valera, T. S. ; Sakai, R. ; Coelho, A. C. Vieira ; Wiebeck, H. ; Toffli, S. M. - Reuse of glass-cutting waste as a filler in polymerised glass

engineering. In: BRAZILIAN CONGRESS OF CERAMICS, 44, 2000, São Pedro. Proceedings São Paulo: ABCERAM, (2000), 404, 01-08.

Varela, M.L.; Vale, S.A.; Do Nascimento, R.M.; Paskocimas, C.A.; Formiga, F. L. - Utilisation of waste from the construction chain and the kaolin processing industry in the production of ceramic flooring. XXI ENTMME - Natal-RN, November 2005.

Varela, M.L.; Do Nascimento, R.M.; Martinelli,A. E.; Hotza, D.; Melo, D. M. A.; Melo, M. A. F. - Optimisation of a methodology for rational mineralogical analysis of clay minerals. - Revista Cerâmica Vol. 51, p. 388, Oct/Nov/Dec, 2005.

Veleda, V.; Santos, I. S. S.; Kazmierczak, C. S. - Use of Water Treatment Plant Sludge to Produce Ceramic Blocks, 49th Brazilian Ceramics Congress, 2005.

Vieira, C. M. F.; Hygina, S. F.; Monteiro, S. N. - Evaluation of Red Ceramic Drying Using the Bigot Curve. - Cerâmica industrial pág. 42-46, Industrial Ceramics, 8 (1) March/April, 2003.

Vieira, C.M.F.; Teixeira, S.S.; Monteiro, S.N- Effect of Firing Temperature on the Properties of Red Ceramics Incorporated with Chamotte. 49th Brazilian Ceramics Congress, 2005.

Vieira, C.M.F.; Monteiro, S.N.; Dos Santos Júnior, E.L.; - Intorne, S.C. - Incorporation of Steel Slag in Red Ceramics.49th Brazilian Ceramics Congress, 2005.

Wender, A. A.; Baldo, J. B. - The Potential Use of a Clay Residue in the Manufacture of Ceramic Tiles Part I - Characterisation, 2001

. - The Potential Use of a Clay Residue in the Manufacture of Ceramic Tiles Part II - Characterisation

Wild, S.; Sabir, B.B.; Khatib, J.M. - Factors influencing strength development of concrete containing silica fume. Cement and Concrete Research, New York, v. 25, n. 7, p. 1567-1580, Oct. 1995.

Wild, S.; Khatib, J.M.; Jones, - A. Relative strength, pozzolanic activity and cement hydration in superplasticised metakaolin concrete. Cement and Concrete Research, New York, v. 26, n. 10, p. 1537-1544, Oct. 1996.

Printed by Books on Demand GmbH, Norderstedt / Germany